Exploring European Frontiers

Also by Brian Dolan

MALTHUS, MEDICINE AND MORALITY: 'Malthusianism' after 1798 (*editor*)
SCIENCE UNBOUND: Geography, Space and Discipline (*editor*)

Exploring European Frontiers

British Travellers in the Age of Enlightenment

Brian Dolan
Research Lecturer
Wellcome Unit for the History of Medicine
University of East Anglia
Norwich

First published in Great Britain 2000 by
MACMILLAN PRESS LTD
Houndmills, Basingstoke, Hampshire RG21 6XS and London
Companies and representatives throughout the world

A catalogue record for this book is available from the British Library.

ISBN 0–333–78987–3 ✓

 First published in the United States of America 2000 by
ST. MARTIN'S PRESS, INC.,
Scholarly and Reference Division,
175 Fifth Avenue, New York, N.Y. 10010

ISBN 0–312–23051–6

Library of Congress Cataloging-in-Publication Data
Dolan, Brian.
Exploring European frontiers : British travellers in the age of Enlightenment /
Brian Dolan.
    p.   cm.
Includes bibliographical references and index.
ISBN 0–312–23051–6 (cloth)
1. Travelers—Great Britain—History—18th century.   2. Europe—Description
and travel.   I. Title

D907 .D665   1999
914.04'25—dc21
                                                                    99–051808

This book is printed on paper suitable for recycling and made from fully managed and sustained
forest sources.

10   9   8   7   6   5   4   3
09   08   07   06   05   04   03   02   01   00

Printed and bound in Great Britain by
Antony Rowe Ltd, Chippenham, Wiltshire

For my mother, Janet, and in the memory of my father, Earl, who taught me how to travel

Contents

List of Plates

Plate 1. 'Norr-Malm Square in Stockholm'. Royal Palace shown at the front and the Opera House to the left. From E.D. Clarke, *Travels in various countries of Europe, Asia, and Africa*, Vol. 3 (London, 1819) (courtesy of the Wellcome Institute Library, London).

Plate 2. 'Portrait of the Regicide Ankarström' (tied to scaffold, assassin of Gustav III). From E.D. Clarke, *Travels in various countries of Europe, Asia, and Africa*, Vol. 3 (London, 1819) (courtesy of the Wellcome Institute Library, London).

Plate 3. 'Mode of Exhibiting the Bodies of Criminals in Sweden' (engraving of a decapitated body on top of a tree trunk; the head and right hand are on separate trees). From E.D. Clarke, *Travels in various countries of Europe, Asia, and Africa*, Vol. 3 (London, 1819) (courtesy of the Wellcome Institute Library, London).

Plate 4. 'View of the great Mine of Copper at Fahlun in Dalecarlia'. From E.D. Clarke, *Travels in various countries of Europe, Asia, and Africa*, Vol. 3, Part the second (London, 1823) (courtesy of the Wellcome Institute Library, London).

Plate 5. 'Laplanders, having prepared their Winter Tents'. From E.D. Clarke, *Travels in various countries of Europe, Asia, and Africa*, Vol. 3 (London, 1819) (courtesy of the Wellcome Institute Library, London).

Plate 6. 'Mode of forcing a Passage through the Ice, when the Sea is not sufficiently Frozen to sustain the weight of the Human Body'. (Clarke and his fellow traveller Cripps off the coast of Finland, in 49 degrees Fahrenheit below zero weather, watching their guides break up the ice with their oars.) From E.D. Clarke, *Travels in various countries of Europe, Asia, and Africa*, Vol. 3, part the second (London, 1823) (courtesy of the Wellcome Institute Library, London).

Plate 7. 'View of the interior of the Kremlin showing the ancient Palace of the Tsars and the first place of Christian Worship in Moscow'. From E.D. Clarke, *Travels in various countries of Europe,*

Acknowledgements

Travel is a fascinating subject to write about, but periodically I drifted off course, lost my way and wondered if I would ever successfully complete the journey. A number of people helped me along the way and tried to keep me on the right path. I would like to thank Will Ashworth, Henry Atmore, Jim Bohun and Karl Galle for reading drafts of the manuscript. As always, Bob Hatch, who not only patiently listened to my evolving ideas about the book, but downloaded and read various draft chapters sent on e-mail, has provided many stimulating comments. I've benefited from the expert advice of Sverker Sörlin (particularly while I worked at Umeå University) and Anders Lundgren on aspects of eighteenth-century Sweden, and Professor Rod McGrew who kindly drudged through the Russia chapter. I thank everyone for their suggestions, comments and cautions. Naturally, I exonerate them from any remaining (or reintroduced) errors I have made.

I would also like to express my gratitude to the staffs at the various institutions where I conducted research and worked on revisions of the manuscript, including Cambridge University Library, the British Library and the London Library. I am especially grateful to the staff and photographic department at the Wellcome Institute for the History of Medicine in London for supplying volume upon volume of different editions of travelogues for me to chase up references, and for permission to use illustrations from them. I should also like to thank Aruna Vasudevan, and Ruth Willats for saving me from many errors.

Finally, I offer a special thanks to Dorothy Porter for her faith and encouragement, and with whom I enjoy travelling through life so much.

Death of the Grand Tour

For what purpose do you travel? Is it to associate promiscuously with adventurers? – to be immured in gaming-houses? – to be seen all morning at the billiard table; and all the evening intoxicated; or, at the faro bank? – to become the object of contemptuous ridicule in every country you visit? Is it for this Albion pours forth her sons upon foreign ground; in the vein hope of obtaining ornaments to her senate, honours to her state, understandings enlarged, prejudices corrected, and taste refined?

– Italy, exhausted by a long and successful scrutiny, is unable to supply new gratification, either in art or antiquity. But in other countries, removed from common observation, new fields of enterprise open an extensive prospect of pleasing research; as the desolated shores of Greece; the peaceful islands of the Aegean; the interesting plains of Asia Minor; the lakes, the ruins, and volcanoes of Syria; and the long, hollow valley of Egypt.

These are the countries to which I would invite your attention. Among these scenes I would lead you to rescue from indiscriminate ruin, the marvellous profusion of antiquities which lie scattered in promiscuous devastation, and yield a daily tribute to the wants or superstition of the inhabitants. . . . Instead of being harassed at Rome, by a perpetual cabal of antiquarians and artists, whose intrigues and discord pervade all the avenues of inquiry, and interrupt the progress of your studies, I will strive to withdraw you to those delightful scenes, where imposition has not yet dared to intrude; where, fearless of her snares, you may investigate the ruins of empires, whose inventive genius first produced, and then carried to perfection, those arts, which Rome in the zenith of her glory could only imitate. I invite you to extend the sphere of your ideas, that reflection may cast off the yoke of prejudice, and break the bonds by which custom as enchained the flights of human reason; to walk among the sequestered pillars of Athens, or trace the mystic labours of Egypt upon the pyramids of Memphis; to mark the chisels of Praxiteles and Phidias, among the mouldering fabrics of Greece, or drop a tear to literature over the august ruins of Alexandria . . .

Edward Daniel Clarke, 'Letters on Travel to the Young Nobility of England', in William Otter, *The Life and Remains of Edward Daniel Clarke*, 2 vols (London, 1825), Vol. 1, pp. 246–8

1
Preparing the Course:
The Death of the Grand Tour and
the Making of a Literary Traveller

Identity, history and travel

Across Europe historic communities are re-emerging, fighting for devolution, autonomy and the right to be recognised as politically independent. Yet at the same time, a powerful sense of a shared European identity has prevailed: in 1991 'Ode to Joy' was played at Slovenia's proclamation of independence and the EU flag was waved by anti-Soviet demonstrators in St Petersburg. One is quick to associate these events with rapid political changes over the last decade. Yet the revolutions of 1989/90 which overthrew communism in Eastern Europe – knocking down the Berlin Wall and, in December 1991, dismantling the USSR – complicated a much longer historical process whereby a 'European' identity was being conceptualised, debated and redefined within many different nations. The lifting of the Iron Curtain and the fission of Eastern states redrew political frontiers, redefined cultural boundaries and ended the Cold War dichotomy between West and East which had helped define 'Western' identity since the Second World War. Today, it is more difficult for the British to sustain a separate identity, without a politically and ideologically *different* nation on the other side of the Continent.

It was also in December 1991 that member states of the European Community agreed through the Maastricht Treaty on a course for future integration. With the idea of creating a new, 'unified' Europe emerged a schedule according to which an era of peaceful co-operation and closer political and economic union could be forged.

3

The creation of a single European market was planned and the agency of the European Union was established. Barriers were broken down, free trade permitted amongst members of the Community and legal restraints on travel – imposed after the Second World War – were relaxed.

But when we review attempts to forge a new Europe, a pan-European nationalism or a unified community growing towards a globalisation of trade and political policy, some striking paradoxes are uncovered. The definition of what might constitute a 'European identity' has been plagued by problems. Ironically, attempts to strengthen Europe through unification have been somewhat disturbed by the collapse of the Eastern states against whom the Western powers had initially desired to consolidate their strength. In post-revolutionary Europe, it becomes politically sensitive to feign a highly selective federated 'super-Europe', which sits snugly against those 'outside' the Community, the marginal or peripheral peoples who have recently rediscovered their historical, regional identities. Never defined easily through geographical reference, Europe today is becoming synonymous with concepts of homogenisation, political conformity and economic standardisation.[1]

A number of questions can be asked concerning the present condition. In what way does the political rhetoric of economic unification lend itself to new conceptions of European identity? Is there such a concept that can apply to groups less familiar to our political, economic and historical narratives? Does sharing such an identity simply mean trade and travel agreements? The quest for a unified, inclusive European identity is guided by arbitrary criteria which demarcate geographical boundaries, set 'membership' standards and often ignore regional distinctiveness and national-historical consciousness. Struggling with concerns and problems over thinking of what it means to be European, however, is not new to the late twentieth century. British perspectives on these issues have matured over hundreds of years of travel and foreign encounters, and, since the eighteenth century, we find similar questions to our own raised in travel narratives and contemporary public debate.

While always arbitrary in one way or another, concerns over who counts as European – geographically or historically – have an intellectual, political and artistic history. But the level of concern to define 'the European' seems to be proportionate to the degree of political power emanating from central Europe. To whom did it matter who or what was considered European? Questions of identity mattered most to those who never questioned their own status as such; to those at the 'margins' – those whose identity was in question – these concerns or judgements were irrelevant, if even acknowledged. Problems of European identity have developed as Eurocentric problems, contemplated in similar ways by scholars from, for example, Britain, France and Germany. How, then, does exploring the ways that the British have characterised others around the Continent relate to broader themes in European history? One significant link is that the history of British perceptions of those who counted as European is also a history of the rationalisation of the classification of populations. Through the development of the sciences of demography, anthropology, philology and natural history during the Enlightenment, taxonomies of humankind placed conceptions of 'Europeanness' on new epistemological bases. From the eighteenth-century philologist Sir William Jones to the nineteenth-century social anthropologist James Cowles Prichard, individuals from Britain with a range of interests used the social-scientific route to the classification of human kinds to seek new evidence for their accounts of European ancestry.

In twentieth-century cultural studies, Western society has been criticised for its presumptuous naiveté towards other cultures, for its ethnocentricity and for its cartographic biases and Eurocentric imagery. The 'Other' in Western civilisation has a long history of being pushed and prodded, explored and colonised, misrepresented and appropriated. While such critiques have forged new levels of cultural awareness and cross-cultural communication, we have also begun to see ways that cross-cultural encounters have been more dynamic and interactive than previously portrayed; they cannot solely be represented as a monolithic, imperialist conquest. Travellers in foreign territories witnessed different ways of life, lived in strange and motley conditions and their travel diaries often revealed their anxieties. Not only recording the contours of the landscape, travel writers ineluctably left traces of their psychological journeys. As a result of what can be considered the 'foreign effect', encounters away from home – given enough time and through the active interests of enough people – not only transformed the foreign into the familiar, but provided new perceptions of and reflections on life at home. As Ralph Waldo Emerson wrote of his countrymen travellers, 'We go to Europe to become Americanised.'[2] Through the dialectics of travel and encounters with the Other, both self and foreign identities have been, and are continuously, reconstructed.

Questions of how international identities have been historically treated are further relevant because they allow us to see how cultural relativity has grown to be part of the treatment of foreign, as well as one's own, society. Whether referring to present concerns over human rights and environmentalism or historical concerns over imperial expansion, the distribution of disease, or rights to 'citizenship', different nations have used cultural comparisons to distinguish the progressive society from the barbaric, the civilised from the uncivilised, the modern from the 'traditional'. These categories, like all classification systems, have always had problematic boundaries. But through travel and the uses of Enlightenment 'sciences of man' to inspect foreign frontiers, strides were made to map the margins of the historical and scientific classification of populations – 'primitive' or 'enlightened', within a 'European' or 'extra-European' domain.[3]

Studies of how scholars at different times and in different nations have represented the Other – foreign, distant or culturally unfamiliar – have become a prominent feature of literary and historical scholarship since the publication of Edward Said's *Orientalism* (1978). Said explored how comparative linguistics, anthropology, travel literature and museum artefacts were used by Western Orientalists to construct representations of the exotic 'Orient', the term being derived from travellers' references to the land lying eastward, 'the place of the rising sun'.[4] The literary traveller did not just move through foreign territories, but recognised a rich interplay between the diverse, natural geography and the different levels of civility of the people who occupied these territories.

From Enlightenment thinkers such as the German jurist and historian Samuel Pufendorf (1632–94) to the Scottish conjectural historians, the development of human civilisation was conceived as being an historical process that followed lines of progressive refinement of social courtesies and increasing cultivation of arts and sciences.[5] In *The Civilizing Process*, Norbert Elias described how different degrees of civility were marked by changes in the standard of behaviour in individuals of a society and how this reflected changes in a society's psychical make-up. He charted the various ways that societies at different stages in the civilising process have appeared to Westerners as 'younger' or primitive, and others 'older' or maturer.[6] Within an historical-sociological model, developed in the late Enlightenment, the measure of civility was used as a way of determining who was advanced enough to be considered part of modern Europe.

But while such measures were deployed to evaluate the status of many exotic groups around the world, rarely was it thought necessary to consider the measure or status of European civilisation. However, the ends and limits to cross-cultural comparison somewhere met the European frontier, and defining these boundaries became an issue relevant to many eighteenth-century political and scientific concerns.

Crises of identity precede as well as follow periods of revolution. The political critiques which followed the revolutions of 1989 present illuminating parallels to the intellectual enterprise of comparative cultural analysis that emerged in the wake of the French Revolution in 1789. This study examines how the activities of a variety of British travellers to the Continent informed cultural critiques of European nationalism and historical identities in the years preceding the Congress of Vienna and the establishment of the 'Concert of Europe' in 1815.

The making of a literary traveller

Perusing literary reviews or the shelves of most bookshops reveals the present popularity of travel writing. But this is not a new phenomenon. Since the eighteenth century, travel has been considered romantic, heroic, pioneering, exploratory and militaristic. Thinking of all the different reasons for travel quickly presents a vaudeville list: migration, political exile, military exercise, missionary work, pilgrimage, trade, exploration, improvement of health, ambassadorial duty – the list goes on. It has never been the sole privilege of the rich, nor merely for leisure or 'gentlemanly education'. Travel had the potential to enlighten the individual as much as it endangered travellers' lives; it was capable of leading to the appropriation of indigenous artefacts as much as a new understanding of distant cultures.[7] Travel narratives relayed the nuances, quirky events, subtle details, dangerous escapades and reflective moments to readers who thumbed through their pages in London coffee-houses or strolled along the shelves of lending libraries.

Historians of British culture have recently presented us with rich accounts of the fashioning of its political, social and commercial life in the eighteenth century. In *Britons*, for example, Linda Colley showed how constituents of England, Wales and Scotland, largely through collective military and religious opposition to hostile European neighbours, forged a new national identity, and indeed a spirit of patriotism – a 'Britishness' – during the eighteenth century. More recently, John Brewer's *Pleasures of the Imagination* offered a different perspective, and explained how a burgeoning commercial culture transformed the market for aesthetic and literary enterprise, marking the appearance of a modern 'high culture' in British society. However, what has been lacking is a more focused account of how the British grew familiar with and defined 'modern Europe' – ways they explored different neighbours' civilised status and evaluated their cultural achievements.[8]

Understandably, British relations with France (at least among the patrician classes) have been a key focal point in historical accounts of how British identity and nationalism emerged during the eighteenth century. Unquestionably, *beaux arts, belles lettres*, French design and the writings of the *philosophes* helped shape certain aspects of British taste, fashion and philosophical criticism; France remained influential, even if in an opposing way, throughout the French Revolution and the Napoleonic Wars. If, as is further discussed below, identity is considered to be defined relative to, and in reaction to, the 'Other', then France – the closest continental neighbour, imperial rival and land of religious antagonism – was certainly a nation with an inescapable affinity.

But what about other continental countries? Did British travellers to the frozen north of Europe, the peasant-warrior land of the Cossacks or the sun-baked Levant reflect on their Britishness, or, indeed, on the idea of a shared *European* identity? Looking at the activities of travellers to places on the European periphery provides us with examples of how the 'foreign effect' might have been brought to bear on matters of taste, national identity and even broader cultural conceptions of what constituted 'Europeanness'. Britishness might have been defined relative to perceptions of different European national identities, but did the British became more consciously part of broader European culture when confronted with scenes of the savage, the uncivilised and the barbaric outside the (not so well-defined) geographical boundaries of 'Europe'? Little is known of how British travellers constructed images and relayed accounts of the diverse groups of the people that lived in what were considered the borderlands of European civilisation. In this study, we begin to see how the British did in part identify with cultural values from elsewhere in Europe and shared analytical tools to evaluate the civilising process. But we also see that more refined attempts to redraw the map of civilisation to underscore what constituted cosmopolitanism or enlightenment civility made the classification of 'the' European more problematic.

On the other hand, much has been written about the heroic travels of the British to distant and exotic lands in the South Pacific or across the Atlantic to the 'New World'. Mungo Parks' eighteenth-century explorations of the 'dark continent' of Africa and the voyages of Captain Cook between 1767 and 1779, partly in the company of Joseph Banks, who later became the President of the Royal Society (the oldest scientific institution in Britain), have been areas of research well traversed by historians of science, economics and anthropology. We have thus become aware of the curious nature of much that went on during these trips: Joseph Banks leaving Tahiti with a Tahitian boy as a souvenir whom he planned to keep on display the way 'some of my neighbours do lions and tygers'; the descriptions by many travellers to the East Indies or Africa who reported finding giants, dwarfs and sea-monsters, or relayed countless other tales of encountering 'wild men, hairy men, bearing tails . . . midway between apes and us'. This was, after all, the age that created mermaid myths and first confronted chimps and orang-outangs.[9]

In his study of *European Encounters with the New World*, Anthony Pagden described how 'the European' 'recognized, confronted, and explained' the savage and barbarian peoples who were 'discovered' on early modern voyages to the Americas. Sacrificing what was even in eighteenth-century standards acceptable ethnographic practices, the European travellers interpreted the bizarre behaviour and rituals of the Amerindians in such a way as to make them recognisable to Europeans. By distorting or ignoring the native meaning of the actions or intentions of the indigenous population, the visitors used familiar terms and their own cultural resources to make sense of what they witnessed.[10] Such attempts to assimilate the unfamiliar into the realm of the familiar were used to help absorb the shock of the different in distant lands.

But those adventures were far from home; the sheer number of dissimilarities, both geographical and social, already required readers of these tales to take a long leap of imagination.[11] But what would have been the reaction to encounters with 'wild men' if they were found closer to home? One British traveller who ventured to northern Swedish territories described the 'wild Laplanders' he encountered as 'wretched pigmies', 'out of the order of nature'. A Scottish physician who travelled to the southern provinces of Russia was startled by the appearance of the 'Kalmuck' Tartars, who were bald and had small black eyes, wore loose sheep skins and were armed with bows and arrows. They seemed to confirm Edward Gibbon's judgement that 'the wild people who dwelt or wandered in the plains of Russia' existed half-way between barbarism and civility.[12] The language describing these people was not so far removed from the language describing those 'discovered' half-way round the world, but neither were these people *geographically* far removed from central Europe. To many travellers and social commentators this was disturbing. After all, these observations were made in the age that coined the term 'civilisation', with Europe at its hub. For Europeans travelling within Europe, it was the task of the modern traveller to familiarise those at the centre with life at the margins of civilisation, rather than force the exotic into preconceived norms of European behaviour.

It was one thing to construct classification systems embracing the pluralities of human species when the 'specimens' were collected in remote geographical areas. In different places, climate may have caused 'degeneration' of the archetypal human or God may have created different beings as links in the 'Great Chain of Being' which connected the most simple to the most complex forms of life. But it was quite another thing to direct such questions about human degeneration and diversity towards the European continent. Encounters with characteristically distinct groups on one's doorstep broadened conceptions of the diversity of people and ways of life around (sometimes arguably *within*) Europe, as well as enabled readers of travel narratives to form conceptions of the historical, topographical and cultural limits of European civilisation. Such inquiries into the conditions of 'human nature' or racial comparisons became more sensitive when in such intimate proximity to the European metropolitanate who prided themselves on being at the historic cutting edge of the civilising process. Thus, travellers' activities and the popular genre of travel literature made comparisons between life at home and life at the edges of Europe an intriguing, if also controversial, intellectual issue.[13]

A swelling area of research in cross-cultural studies shifted the direction of all sorts of scholarship in the eighteenth century. Comparative methods of study became prominent, and were applied to natural history, the environment and society, and extended to new areas of investigation including comparative anatomy, philology and political economy. The human sciences (including archaeology, ethnology and anthropology), the biological sciences (such as physiology and physiognomy) and the social sciences (such as demography and economics) were slowly evolving into specialist disciplines. As Adam Ferguson (sometimes called the 'father' of British sociology) argued in his *Principles of Moral and Political Science* (1792), 'external accommodations, diversity of manners, and forms of policy' affected humankind in different ways, leading to varied circumstances in the organisation of social and political life in different national contexts.[14] This invited comparative studies between the various 'sciences of man'. Travel, and travel literature, lay at the heart of this enterprise.

In Britain, various eighteenth-century institutions provided patronage and forums for discussion, debate and public pedagogy in the arts and sciences: the Society of Antiquaries in London (f. 1707, royal charter 1751), the Society of Dilettanti (f. 1734), the British Museum (f. 1753), the Royal Academy of Arts (f. 1768). The 'modern', turn-of-the-century travellers who will be discussed here were not only products of a commercial culture which legitimised professional literary and philosophical pursuits, but shared many concerns to compile and classify specific forms of information about human society and material culture. We have grown familiar with the political critiques of civil society by Ferguson and John Millar, the Orientalist scholarship of William Jones and the influence of the French *philosophes* or the political physiognomy of Montesquieu on Enlightenment thought. It is the aim of the chapters that follow to add more voices to this chorus, and show how others – doctors, dons and diplomats – helped create a heightened awareness for the scientific and literary study of particular aspects of human culture.

Travel books, being part and parcel of allied subjects such as history and geography, were among the most frequently read books in Georgian Britain.[15] Travel writers became central figures in the growing profession of letters and benefited enormously from the increasing encouragement and demand brought forth from a burgeoning literary culture. England may have become a 'nation of shopkeepers' in the eighteenth century, but, as Samuel Johnson – the enthusiastic patron of literary endeavours – remarked, it was also becoming a 'nation of readers'.[16] During the last quarter of the eighteenth century a number of authors had penned popular and influential works on European travel and history. Whether literature on the Orient, such as Lady Mary Wortley Montagu's *Turkish Embassy Letters* or histories of Rome or Greece by the likes of Oliver

Goldsmith or the more scholarly Edward Gibbon, the Continent provided inexhaustible themes for men and women of letters.

Growing numbers from Britain who embarked on the Grand Tour complemented the increasing amount of literature about Europe that rolled off the printing presses. In 1785 Gibbon was allegedly informed that 40,000 Englishmen were on the Continent; thirty years later, following a dip in foreign travel, the Battle of Waterloo and Napoleon's defeat, the number of British tourists travelling abroad was cited as then having surpassed 45,000 a year. As has been well documented, France was the first, and sometimes the only, stop on the traditional 'Grand Tour'. From there, by far the most well-trodden track was through Dijon, Milan and Florence, finally ending in Naples, where countless tourists, artists, sculptors and writers gathered around the fashionable circles entertained by the British envoy, Sir William Hamilton. Recent research shows that, in the last two decades of the eighteenth century, the largest percentage of tourists came from the professional classes, many of whom were authors – more of the Smolletts, Goldsmiths and Gibbons of eighteenth-century Britain. The role the Grand Tour was thought to play in the education of English gentlemen helps explain a predisposition to incorporate French affairs and fashion into the repertoire of polite conversation, again supporting Colley's emphasis on Franco–British relations.[17] For many contemporary *literati*, the Grand Tour was largely synonymous with European travel, a view that has been rather uncritically propagated in current historiography. Just as there were alternative forms and functions of travel, so we must begin to provide alternative historiographic models in exploring this subject.

Between the deaths of Samuel Johnson in 1784 and Sir William Hamilton in 1803, the Grand Tour of the long eighteenth century had also died. Johnson had once been quoted as saying:

> the grand object of travelling is to see the shores of the Mediterranean. On those shores were the four great empires of the world; the Assyrian, the Persian, Greek, and Roman. All our religion, almost all our laws, almost all our arts, almost all that sets us above savages, has come to us from the shores of the Mediterranean.[18]

But by the turn of the eighteenth century, revolution and war in France had prompted new concerns over arts and empire, and had redirected attention to other European fields.

One can observe that whereas Gibbon had invented ancient Europe, many turn-of-the-century British travel writers were concerned to illuminate the modern. Enough of marvelling over the past; it was declared time to contemplate and map the current European state of affairs. In an age of revolution, accounts of fallen empires gave way to contemporary political commentary and attention to the reconstruction of modern empire. Touring countries where military alliances were forged in efforts to defeat Napoleon also meant travelling over new political geography, witnessing other forms of government rule and different methods of maintaining order and discipline. The lessons of history had raised new questions about modern Europe: Who had inherited the principles of proper government? Was evidence to be found of a laxity of public morals or the over-indulgence in the wealth of nations by unprincipled rulers (cited by Gibbon as causes for the fall of empires)? How many different parts of Europe could claim to have inherited the traits of civilised society from the venerable shores of the Mediterranean?

Whereas France had previously been the obligatory passage point for Grand Tourists and élite *entrée* into fashionable society, by the end of the eighteenth century it was a land off limits to the literary traveller. The effect of political turmoil abroad on the profession of letters was profound. Especially heedful of political anxieties after the events of the French Revolution and the Terror, a number of writers explored post-revolutionary meanings of citizenship, patriotism, nationalism, war, disease and racial differences. These included the Romantic poets, such as Wordsworth and Coleridge or Byron and Shelley – some extolling the virtues of religious orthodoxy and political loyalties, others musing about the consequences of imperial expansion and British penetration of the exotic, erotic and outlandish East.[19] But they also included travel writers who were concerned to address anxieties of empire as well as innovations of modern civilisation. So, through an analysis of travel literature, we begin to see more clearly some ways that British cosmopolitan culture became Eurocentric and how novel representations of European culture were packaged for the British public.

British activities in imperial frontiers at the end of the eighteenth century set new agendas for the literary traveller. While domestic tourism flourished in Britain during this time, the few who braved finding alternative inroads to Europe journeyed into territories little known at home and found themselves becoming path-finders for future travellers. Empire, civilisation and principles of government were themes writ large by these travellers who helped transform public awareness and brought these issues into the arena of public debate. Many of these writers were entrepreneurs and liberal professionals who took advantage of the new opportunities in publishing provided by newspapers, periodicals and pamphlet literature. This contributed to the vibrancy of a 'public sphere', a forum for social exchange that gradually subverted the privileged literary patronage of aristocratic court culture.[20] Often we find that ephemeral commentary in these widely circulated, less expensive media of public debate drew on information laid out in hefty quarto volumes of travel narratives.

Some of the travellers discussed in this book who ventured from a stimulating literary culture to the unsettled European climate were university tutors who came from the ranks of the 'middling sorts', a class later to be called bourgeois. As students, they were the sizars and servitors, the Oxbridge achievers with few social privileges who earned their keep by performing menial duties around their College.[21] This was far from being 'enrolled' but exempt from residing at the University or sitting exams, as were the sons of the aristocracy. After graduation, many became tutors who earned money by supervising the aristocratic students' travels, and then by collecting fees from others who attended their lectures after their return to England.[22]

One travelling tutor-turned-writer in particular who acts as a 'travel guide' for us in moving around European frontiers was decidedly ambitious in his attempts to reinvent the meaning of European travel and travel writing at the end of the eighteenth century. The Cambridge graduate Edward Daniel Clarke (1769–1822) was exemplary of the new, modern European traveller. For over a decade he travelled extensively through Britain and Europe, and published volume after volume of travel narratives throughout another decade while lecturing to Cambridge students about the benefits of travel for the promotion of the arts and sciences in modern society. Despite his notable legacy as a traveller, travel writer and pedagogue, he has received scant historical attention and is thus worth introducing here in some detail.

Clarke received his Bachelor of Arts from Jesus College in 1790, at the age of twenty-one. In the Senate House examinations he obtained an unflattering rank of Third Junior Optime – a low pass – which suggests he was capable, if not gifted, in performing Newtonian and Euclidean proofs, and less than expert in classical languages and moral philosophy. Nevertheless, college connections and family friends recommended him as an adequate tutor to super-intend the education of high-born students.

Clarke caught the cusp of the customary Grand Tour: the role of the 'tutor' was not to prepare patricians for the university exams, but to guide them into their fashionable lifestyles. To be genteel and 'Cambridge-bound' meant paying a stipend to a don who would organise the logistics of a trip around the country or the Conti-nent. Throughout the 1790s he escorted different students on tours of Britain and southern Europe – twice trekking through France down to Italy.[23] In 1798, after settling back in Cambridge on a College Fellowship, he was contacted by the family of John Marten Cripps, a nineteen-year-old who had just inherited a handsome income of £2,500 a year, and who sought his services to supervise yet another European tour. By then, Clarke indeed felt as if 'a map of the world was painted on the awning of my cradle'.[24]

Clarke negotiated a deal for Cripps to be 'entered' as a Fellow-Commoner (a privileged social position among Cambridge students) at Jesus College, that Clarke would receive £300 per annum stipend and that the incurred expenses of the trip would be settled by Cripps. Clarke was responsible for organising the route and managing the details, such as obtaining letters of introduction and credit, pro-curing translators, and hiring horses and guards. Insights on organising the trip were gleaned by examining as many previously published British or foreign travel narratives as were available. Nothing was extraordinary about these arrangements except that this time, Clarke intended to make an entirely different journey and to explore parts of the European map he had not previously unfolded.

Encounters with Clarke

Cripps was eager to travel and able to fund the way to whatever frontier Clarke desired. The cultural and political climates were ripe for a new initiative from the tutor with a decade's travelling experi-ence. In order to circumvent the 'distracted state of public affairs' in France, they headed for the north of Europe to travel around the Baltic. 'A curiosity to visit the Eastern boundaries of Europe [was] naturally excited by the circumstance of their situation,' Clarke later explained to the readers of his *Travels*, and they were keen to visit countries that had been 'rarely traversed by any literary traveller, and a little noticed either in antient or in modern history'.[25]

Although they initially planned only to visit Scandinavia, Cripps endeavoured to keep apace with his tutor, and Clarke's ambitions to cover more ground and push further into European hinterlands never eased. Their entire trip was to prove extensive and rewarding beyond their most sanguine expectations. They would explore and collect what few others in Britain had ever seen and fewer still had written about. They went from the extreme north (above the Arctic Circle), through Russia and Tartary, to the Holy Land, northern Africa and Greece. It took them two and a half years to complete their journey, and altogether they trekked thousands of miles over little-explored lands.

Clarke was alert to the peculiarities of place that surrounded him throughout the journey. He was mindful to record his observations of the customs, habits and manners of the people, to transcribe samples of script, and to draw or collect any curious artefact he or his fellow traveller happened upon. Backed with the purse of his patrician tutee, he was able to purchase and ship home manuscripts, coins, precious stones, statues, natural history specimens, and the like, as suited his intellectual interests. His collection grew with an eye to feed a swelling area of research in cross-cultural studies.

At the time of Clarke and Cripps' trip, other travellers had provided only glimpses of the northern and eastern European nations. Just three years before they embarked, Mary Wollstonecraft had published her *Letters written during a short residence in Sweden*, which propagated romantic imagery of picturesque Scandinavian landscapes. William Coxe's *Travels in Poland and Russia* was in its third edition (1792), which provided welcome observations on the Russian empire which helped satisfy curiosity in part aroused by Peter the Great's British tour one hundred years earlier. Not until the 1770s did William Jones' philological pursuits provide a scientific study of Indian culture (while also helping to shape a new discipline in the study of comparative languages). And the study of Greek artefacts was introduced to British art and archaeology through James Stuart and Nicholas Revett's lavishly illustrated *The Antiquities of Athens* (see chapter 4).

The stories told by travel writers earlier in the eighteenth century were enough to attract general interest in the far-flung regions of the Continent, but Clarke – to focus on one influential spokesperson – thought it due time to supplement general impressions with a detailed, analytic account. Particular places were, of course, targeted as stopping points throughout their journey, such as the metropolises and famous universities, but they were few. Rather, they moved according to local recommendation and folklore, spontaneous whims of curiosity and investigative pursuits for relics and antiquities, the 'trophies of travels' which would adorn museums, libraries, cabinets and lecture theatres on their return.

Further, Clarke was conscious of the intellectual demands that one should bear when navigating the boundaries of what was 'Oriental' and 'Occidental' (Eastern and Western), mapped and unmapped, civilised and barbaric. From the Age of Discovery sprang a powerful awareness of the diversity of nature's resources and a desire to seek the unknown. Clarke combined a spirit for discovery with the tools of rational exploration and explanation. A child of the Enlightenment, he used a modern rationale – drawn from natural philosophy, philology, geography, and so forth – to guide new perspectives on what constituted a European identity. Since he travelled at the very end of the century, he was able to draw on and refine earlier travellers' observations.

While the use of historical models to 'map' the civilising process were widely used by various eighteenth-century writers across Europe to help demarcate developed or 'modern' civilisation from those 'underdeveloped', other kinds of models – more anthropological than historical – were also used, notably by Clarke. In this sense, rather than thinking of Clarke as an early historical sociologist concerned to map the civilising process, we might think of him as a traveller concerned to trace cultural commensurability, to chart new territory and familiarise through description different and distant parts of the continental map.

His account of European frontiers (along with others') was not just another story, but an unfolding of human relations, reflections on customs and regional identities, and an identification of common characteristics amongst people whether savage or civilised. His observations of the customs, habits and manners of people from the Arctic Circle to the Levant reveal ways in which the traveller came to terms with different surroundings at the moment of foreign encounter. What is crucial to his account is that Clarke recognised the 'Other' as being at home. In such a narrative, the traveller becomes the 'Other', the outsider; he appears vulnerable – dislocated from the familiar provinces of central Europe.

Arguably, the late eighteenth century witnessed a significant shift in styles of travel writing and cultural analysis. Previously in travel narratives the writer was the subject, and here the personal voice and self-conscious reflections marked the genre. In the late eighteenth century, when authors wrote about their travels to the 'uncivilised' parts, their self-conscious status was inverted from being part of secure cloisters of polite society to being exposed foreign bodies themselves susceptible to probing eyes, physical attack and disease. Travelling off the beaten track, Clarke (and other travellers) confronted and reported on the unknown which was the focus of attention. The 'eye' of the traveller, which in one genre of travel literature offered readers a glimpse of adventure and heroic survival, was replaced with more acute descriptions of the culturally distinct.[26]

Clarke was one author to write with new focus, but he was not alone in attempting to introduce commentary on the character of modern society. He seized the moment and aimed to write a travelogue that would set the pace for future travellers. Later, de Staël writing about Germany, or de Tocqueville on America, gained similar contemporary credit for their travels and observations. But writing in the first two decades of the nineteenth century, Clarke proposed a new perspective on what defined modern, rather than historical, Europe. And here the term 'modern' itself was used with specific meaning.

Following the Renaissance the idea of the 'modern' developed as a presentist mentality: by historians it was used consciously to distinguish an epoch from the remote past (such as the ancients versus moderns debate); it was used in art and architecture to refer to something not obsolete or antiquated; by grammarians to denote differences between contemporary 'modern languages' and ancient, classical languages. But characterisations of something modern were at times more carefully defined. In the late eighteenth century the concept of 'modern' was used as a classificatory category of certain traits or characteristics in human society. To be considered modern, or to be concerned with the 'modern', was not merely to be included with the present era, but to have distinguishable traits or characteristics. The job of defining modern characteristics was assisted by the development of different ways that the human sciences in the eighteenth century studied society.

Travellers were not only conscious of the values that collecting information had in helping define modern society, but through their efforts they helped articulate the notion of a 'modern epoch' (further developed by later theorists such as Ferdinand Tönnies, Auguste Comte and Émile Durkheim). Clarke's contribution to the making of a modern mindset is even recognised in the *Oxford English*

Dictionary, where, under 'modern' and in reference to its meaning as a current historical period, it cites Clarke's distinction between the 'modern Greeks' and the ancient Getae (who were associated in the eighteenth century with the Goths). What was modern was new, and Clarke wrote his travel narratives with the ambition not only to chart European frontiers, but, with the spirit of what I refer to as the *naissance de siècle*, to redefine modern travel and travel writing. It was a spirit that attempted to rejuvenate Enlightenment optimism, shaking loose the shackles of end-of-century gloom, and searching for alternatives to Romantic disillusionment with social development and political pessimism spawned by fears of widespread revolution.

Arguably, in terms of geographical scope and in sales, Clarke's narrative of his European journey proved a more successful publishing enterprise than any other traveller of his period, placing him firmly on the map of Georgian *literati* (the famous artist John Flaxman illustrated a portion of his publications, and his correspondence network included the likes of Lord Aberdeen, Lord Byron and Sir Joseph Banks). No one else covered so much ground or wrote so many pages (nearly 6,000 pages in eleven volumes of octavo text). The first edition of his *Travels* was published between 1810 and 1823, appearing in six, heavy, elegant quarto volumes, richly illustrated with over fifty copper-plate engravings. They were expensive – £23 for the set (just under £10 for subsequent editions printed in eleven octavo volumes) at a time when an artisan family lived on roughly £1 a week – but they were well received. The publisher, Cadell and Davies, was known as the 'aristocracy of the trade', noted for publishing authors including Johnson, Blackstone, Gibbon, Fielding and Hume.[27] By 1848, six editions of Clarke's *Travels* had been issued, octavo volumes from the third edition (1816–24; duodecimal volumes in the sixth). The first quarto volume was translated into French (two editions, 1812 and 1813), German, Swedish and Italian. Further editions were printed in New York. This publishing success clearly ranks them on par with those of other famous travellers (such as Prevost, Pinkerton or Coxe) and armchair travellers cashing in on the demand for pulp reading.

Clarke's *Travels* earned him nearly £7,000 and the reputation as being 'The Traveller' for the next 200 years. The breadth of his accomplishments surpassed the publication of his narrative, to include the collection of 76 boxes containing medieval manuscripts, maps, marbles and minerals, in addition to a two-tonne statue of

Goddess of Agriculture. He is today regarded as one
lanthropists of art (for donating, rather than selling,
..easures to public museums), and the inventor of a
to 'date' ancient monuments scientifically through chemical
analysis. Certain events during his travels associated him with all
different levels of public notice. He travelled to Scandinavia with
his colleague from Cambridge, the controversial historical demo-
grapher Thomas Robert Malthus, and provided some data for Malthus's
infamous *Essay on the Principle of Population* (2nd edition 1803).
In Egypt he assisted a group from the Society of Antiquaries in
London to procure the Rosetta Stone from Napoleon's soldiers after
the British victory over the French in Alexandria. In Athens Clarke
watched Lord Elgin's agents dismantle the Parthenon, and fed Byron
some eyewitness testimony that was footnoted in the poet's *Childe
Harold's Pilgrimage*. More dubious honours of his included his alleged
discovery of Troy, the tombs of Alexander the Great and the math-
ematician Euclid, and a painting by Shakespeare.

His published *Travels*, however, remain his enduring legacy –
monuments to a new spirit of travel and a weighty bestowal to an
increasingly sophisticated genre of travel literature. The present study
does not pretend to provide as much biographical information on
Clarke as he deserves. However, the extent of his travels allows us
to follow him through Europe in order to weave a continuous narrative within a broader geographical context, and in response to a
host of other travellers' observations.

Plotting the course

Focusing on Clarke allows us to use a sole narrative to guide us
through different regions of eighteenth-century European frontiers,
but he was not the only observer of (or commentator on) the places
here discussed. It is best to view him within an historical context
of other travellers' accounts written both before and coeval with
his journey. These not only provide us with diverse and nuanced
accounts, but they provided observations that other travellers con-
firmed, disagreed with or qualified. Because they evolved through
the additions of later travellers – the accumulation of new observations, the publication of successive editions, and so on – travel
accounts are best dealt with diachronically, seen as a growing and
branching body of knowledge about other countries with different
stems written by different authors. Most authors took it for granted

that their readers would be familiar with previous accounts, and some authors borrowed (or blatantly plagiarised) passages from their predecessors, exchanged diaries to gather details and embellished their accounts from a variety of contemporary sources. The more contemporary observations examined the more we sense the richness and diversity in the stock of knowledge available to illuminate aspects of life abroad. But the travel literature also provides a wealth of information and there is only enough space here to provide suggestive glimpses of what was available to contemporary readers. Each traveller's experiences could be examined in more depth and we could learn much about the way they chose to represent foreign lands through more detailed biographical studies of these travellers. Taken as a sampling of the literature in the marketplace, we here only get to taste the different flavours of the European experience dished out by the authors chosen, revealing different interests as well as prejudices. Thus, for example, we can compare Clarke's splenetic hyperbole in attacking Tsar Paul as he quickly passed through St Petersburg in 1800 with Maria Guthrie's fond reflections on Russia when she was resident there for over a decade.

Besides the limitations on the observations that can be discussed, not all authors or types of books that relay foreign encounters could be included. While focusing on eighteenth-century British travellers does allow us to begin to see ways that they grew familiar with certain perceptions of their continental neighbours, other studies examining the abundant writings from the sixteenth, seventeenth and nineteenth centuries would be useful. The same applies for comparing the activities and writings of travellers from other countries around Europe – to relate this study with, for example, Hans Jürgen's *German Travellers in the Age of Enlightenment*. Discussing travellers from Britain alone also precluded an examination of the many available editions of foreign works that were translated into English, an analysis of which would be useful for more detailed studies of British publishing and the literary marketplace. Looking more to the nineteenth century would offer interesting contrasts with new kinds of travel and tourism books, such as Mariana Starke's popular 'Companion Guides' of the 1810s and 1820s to John Murray's famous 'red books', handbooks and phrasebooks published from the 1830s.[28] The writings principally examined here belong to a literary genre connecting travel, topographical and historical narratives, rather than solely descriptive guide-books pointing out peculiar objects in local settings: more of an eighteenth-century *Year in Provence* than *Let's Go*. Choices of organisation go beyond deciding what historical resources should be used. Similar to the ways that travel authors selected their narrative – skipping over space and time – the historian must make choices of inclusion and exclusion. But lessons can be learned from the ways that travel writers presented their stories. The selection of authors represented here embraces the way they selectively represented Europe. In similar ways to travelogues, this book is divided geographically and the narrative is directed by travellers' pursuits. This allows us to see how travellers propagated images of the far reaches of the Continent, in what ways they drew from Enlightenment theories to inspect other populations, how they reflected on the achievements of different societies and how they reached conclusions about the relative status of others in and around Europe. Travel accounts provide the resources and general structure for this book, but were themselves the apparatus with which the British gained information about the Continent.

Each chapter provides an historical underpinning to the comments, criticisms and chronicles offered by the authors discussed. Following closely the ways that European frontiers were represented in travel narratives is an historiographic tool for tapping primary sources that reveal how the late eighteenth-century British themselves conceived of the varied condition of civilisation. This differs from our contemporary accounts, such as the studies in *Enlightenment in National Context* or *Romanticism in National Context*, because in the present study the evaluation of 'enlightenment' in national context is taken from the period.[29] Listening to eighteenth-century authors helps avoid anachronism; there is less risk of imposing modern categories on judgements of taste, political philosophy or cultural condition. The vision of Europe will be unique, sometimes novel and sometimes provocative. The British, as foreigners travelling through unfamiliar territories and living in dissimilar surroundings to their home, noted customs which could not be taken for granted. Those differences, the subtle comparisons and noteworthy nuances of life, helped chart similarities and differences as perceived at the time. These in turn helped define what were considered shared, modern, European values or separate national achievements by the end of the eighteenth century.

The areas of Europe examined in this book – Scandinavia, the Russian territories, Greece and the Levant – each draw out a particular theme pondered by contemporary travellers. They share the general theme of how travellers painted portraits of 'modern' as well as 'historical' Europe. In the northern countries, particularly Sweden, travellers sought to understand what conditions effected a radical change in their political power: the might of the seventeenth-century north had declined into insignificance by the end of the eighteenth. Why? Russia represented the opposite: a rapidly expanding empire, conquering its northern, eastern and southern neighbours with apparent ease. Did they represent a threat to Western powers? Could they be counted amongst the enlightened nations of Europe? In chapter 4 we see how travellers inspected ancient lands, looking for historical preconditions for the making of modernity. Greece – the *locus classicus* for defining the democratic and natural rights for different nations – was also a territorial fighting ground, the site of the imperial frontier for both the French and the British. Classicism became an historical resource for justifying modern social and political structures, and British appropriation of ancient artefacts made British activities in Greece additionally controversial.

The concluding chapter looks at the reception of travellers on their return to Britain. Here we see the diverse museum displays created from travellers' collections of curiosities, whether national galleries such as the British Museum or the Society of Antiquaries, private homes or university collections. But, of course, travellers also re-created worlds through words, and we examine the rationale for publishing travel narratives and look at the increasing popularity of collected and abridged works from a number of travellers for cheaper consumption by the British public. We also see ways that travellers' reports began to feed into tracts in political economy and statistics – the raw data that often guided judgements about the course of future relations between Britain and Europe. But one area where some travellers gained additional 'cultural capital' was in university lecture rooms, where students were provided with firsthand accounts and access to the material culture that they had previously only encountered through texts. Returning to the case-study of Clarke and his new Cambridge professorship, we see how travel could inform new educational ideals. In this example, travel narratives and foreign collections were used as part of pedagogical apparatus to connect disciplines as diverse as natural philosophy, history and classics. This helps us to see how Enlightenment travel provided an alternative ideal for a 'liberal education' (against the emerging 'mathematical' model).

The intersections between politics, the natural and human sciences, and European travel were many. From the Lapps in northern Scandinavia to the Cossacks in Russia, to the contests over imperial reconstruction in the Mediterranean, travellers' narratives publicised different questions about the causes of human diversity, God's providential plan and the progress and condition of civilisation. Sometimes speculation was rash and conjectural, other times based on careful geographical, linguistic and historical comparative analysis. But each in their own way contributed to the eighteenth-century composite portrait of life and customs at the European frontiers.

The Journey Begins

Hamburg, May 28, 1799.

My dear Mother –

We arrived here safe on the 25th, after the most expeditious passage, perhaps, ever known. The captain assured us, that during forty years, he had never reached Hamburg on the third day. Read and determine! We sailed on Thursday at noon. On Friday, at midnight, we passed Helgoland. On Saturday, at half-past six in the morning, we entered Elbe. At half-past ten we arrived at Cuxhaven. Finding a vessel bound for Hamburg, the wind fair, at eleven we started again; and as the sun was setting, at eight in the evening, after a most delightful voyage, we landed in Hamburg. What think you now of our flight? At Cambridge, on the 20th; at Hamburg, on the 25th.

We had few alarms in the passage. Rather a stout gale, as you may suppose by our progress; but not more than the sailors desired. Twice we received signals to hoist our colours; and once we were boarded by the crew of an English hired armed cutter. Otter suffered most in the voyage. Malthus bore it better than anyone. Cripps made a good seaman, being always upon deck.

There are two things which the English expect to receive from Hamburg, viz. news, and hung-beef. The hung-beef I shall keep for our own use among the mountains: the news you are welcome to; and I assure you it is very considerable. Turin is in the hands of the Allies. Naples is taken, &c.

Hamburg is a place of much higher importance than I

had imagined. Her merchants are princes, and their coffers the emporium of the riches of the world. I can buy all sorts of India goods, cambric, Holland, &c. free from any duty. –

We go next to Copenhagen, and from thence along the western side of Sweden, into Norway, to Christiana. We shall then proceed northward as near to the pole as possible. I intend to pass within the arctic, at all events; that for once I may see the sun revolve for twenty-four hours, without setting; and learn what sort of a Rump Parliament they hold in Lapland. We then pass round the north past of the gulf of Bothnia, and afterward cross over to Stockholm and Upsal. Then we visit Finland, and proceed to Petersburg; after which, having letters to Domingo Gonzales, we embark for the moon. Love to all! God bless you!'

Quoted in William Otter, *The Life and Remains of Edward Daniel Clarke*, 2 vols (London, 1825), Vol. I, pp. 451–3

2
Northern Frontier: Scandinavia – The Mismeasure of Modernity and the 'Age of Liberty'

In May 1799, 29-year-old Edward Daniel Clarke, a Cambridge graduate and seasoned 'travelling tutor', set off with his 19-year-old patrician student, John Marten Cripps, on a European tour. Accompanying them were two other members of Jesus College, Cambridge. One, Thomas Robert Malthus, made some of his observations on the early segment of the journey famous when he included them in the second edition of his *Essay on the Principle of Population* (1803). The other, William Otter, would reach the literary world by outliving and eventually writing biographies of both Clarke and Malthus. It would prove to be quite an adventure. For Clarke, it was the culmination of almost a decade of travel throughout Britain and the Continent.

From Yarmouth, a small, merchant-packed port town on the East Anglian coast, the four set sail on the *Diana* and headed for Cuxhaven, the first stop on their European journey. Overall, it was smooth sailing; the British navy was successfully safeguarding the British Channel from Napoleonic despotism, protecting merchant ships and blockading Spanish and French ports. With only a brief delay from an English armed cutter and with the break of a stout gale, the crew arrived in Hamburg, Germany, just three days later. From the outset Clarke kept a close watch on how long each leg of the journey took, working out how to set schedules which efficiently factored time, distance and number of places to visit. Two weeks after they arrived in Germany they were in Copenhagen. The travellers 'have not had above four hours sleep these seven nights passed', Clarke reported in his letters home. It was probably 'that hurry which must attend a traveller' that rendered Clarke indefatigable in the eyes of Otter and Malthus (who went off on their own two weeks later), but which impelled him to traverse the frontiers of Europe over the next three years.[1]

Clarke and Cripps occupied the next six months touring Scandinavia, a region that was relatively unexplored by the English. For four hundred years the national boundaries of the northern countries – Sweden, Norway, Denmark and Finland – had been redrawn and were in constant check. From the Union of Kalmar in 1397, Norway, Denmark and Sweden were politically united, Sweden breaking away from this alliance in 1523 leaving Norway and Denmark, who remained politically joined until 1814 (both countries being ruled from Copenhagen). As a result of the Treaty of Åbo, signed in 1743 after war between Sweden and Russia, most of Finland became Swedish territory.[2] But throughout the eighteenth century different governmental rule in Copenhagen and Stockholm left relations with Russia on the eastern frontier and Germany and France on the southern plagued with territorial tensions.

The state of Scandinavia

Norway and Denmark maintained independent economies and this created internal pressures which increasingly divided the two countries. Trade was one persistent problem: Norway traded with Britain; Denmark traded with Germany, France and North America. A Danish–Russian alliance, strongly backed by Catherine II of Russia, protected Norway from Swedish aggressions. But the expansion of the Russian empire in the eighteenth century was enough of a worry for Sweden. At times, Sweden used military might to keep Russia at bay; at other times, it entered into a tactical alliance with Russia to maintain stable relations, such as during the Seven Years War (1756–63), when Sweden allied with both Russia and France against Prussia. From the late 1740s to the 1760s, the Swedish government met with fair success in their rule which has marked this period as the high point of the Swedish 'Age of Liberty', an epoch of political rule when constitutional and parliamentary reform triumphed over monarchical tyranny.[3] It has recently been argued that the government of this period, in which the phenomena of political parties emerged, bore the hallmarks of what some Swedish historians have called the first 'modern' party system.[4]

British political commentators recognised Sweden's centrality to the geopolitics of the Baltic. A number of Swedish monarchs were selected as heroes of Swedish political emancipation and military prosperity. William Coxe (1747–1828), a Cambridge travelling tutor-turned-historian, who toured the Baltic between 1784 and 1786 with Samuel Whitbread (son of the famous brewer), recorded in his widely read travel narratives a hagiography of the Swedish monarchy.[5] He wrote in adulation of Gustav Adolf II, whose reign (1611–32) he considered the culmination of the great Vasa dynasty, an era when Sweden became a major player in European politics. Adolf built the framework for the Constitution of 1634 which introduced a central *collegium* for the training of civil servants in departments of justice, finance, military and foreign affairs. In promotion of higher education, *Gymnasiums* were founded to provide instruction in classics, philosophy and theology. He endowed the University of Uppsala (f. 1477) with landed property belonging to the House of Vasa, affording them the wealth to expand their faculties and the groundwork to develop into an internationally renowned centre of learning.[6] Coxe summed up the achievements of Gustav Adolf:

> The greatest commander of an age, which abounded in great generals, he stood up as an advocate of liberty and toleration against tyranny, persecution, and bigotry; and laid the first foundation of that equal balance of power which was afterwards settled by the peace of Westphalia.[7]

In 1632 Gustav Adolf was slain on a Saxon battlefield while helping German Protestants defend themselves against the Catholic League. The Peace of Westphalia (1648) ended these wars and was accomplished to no small degree, Coxe suggested, because Gustav Adolf inspired loyal followers to continue his quest that Sweden should be the guarantor of the rights of European Protestants. Even though Gustav Adolf's vision of a liberated Protestant League did not materialise, peace prevailed, and later historians such as Coxe held Gustav Adolf up as a model monarch, a representative of an early modern heroic triumph of religious toleration and liberty. Coxe was a scholar with a special interest in monarchical descent and the history of sovereign rule (perhaps because his father, Dr William Coxe, was physician to the King of England's Household), and during his travels went out of his way to visit royal tombs and catacombs, which also formed much of the subject content of his publications.

The image of Sweden he presented was historical, and he presented a narrative of inherent stability in royal heritage over generations of death, rebirth and remembrance.

Earlier, in 1715, the novelist Daniel Defoe had targeted a different Swedish monarch as the subject of a pamphlet titled *The History of the Wars, of his Present Majesty Charles XII.* Defoe's essay eulogised the achievements of Charles XII, known as the 'warrior king' who reigned from 1697 to 1718. Defoe was impressed with the king's military exploits, writing in his Preface that the subject of his essay:

> is as fruitful of Great Events, as any real history can pretend to, and is Grac'd with as many Glorious Actions, Battles, Sieges, and Gallant Enterprises, Things which make History Pleasant, as well as Profitable, as can be met with in any History of so few Years that is now Extant in the World. The Hero who makes the Superior Figure in this Story, were we to run the Parallel, might vye with the Caesars and Alexanders of Antient Story; He has done Actions that posterity will have room to fable upon, till they make his History Incredible, and turn it into Romance.[8]

For others, there simply was no possibility of a romance. He was a radical king; he broke with constitutional conventions at his coronation by not swearing an oath and by crowning himself. After his death, controversy surrounded who was to become successor, which was finally settled when the Diet elected his younger sister, Ulrica Eleonora. But in 1723 a new constitution was passed which reduced the Crown to a figurehead of the executive government. Thus, this also ended a period of monarchical absolutism which began with Charles XI and inaugurated the 'Age of Liberty'.[9]

Such accounts of Swedish monarchical lineage and of the battles which helped shape the Swedish nation only provided glimpses of Sweden's rise as an imperial power in the seventeenth century. Political changes as well as major demographic and economic shifts during the eighteenth century brought about a reversal of Sweden's imperial fortunes. In the seventeenth century Sweden became a major political power; by the beginning of the nineteenth it was a nation with little European influence. Yet, in juxtaposition to its fall in international politics, Sweden's cultural panorama was enlivened. During the century when the effectiveness of the government apparatus eroded away, Swedish nobles resisted myopic provincialism and embraced Enlightenment pursuits in the arts and sciences so widespread in other parts of Europe.

While accounts of wars and martyrs which shaped the 'northern kingdoms' provided an historical backdrop to the lands under inspection by late eighteenth-century travellers, increasing attention was given to more current events to appear on the Swedish political landscape. When King Adolf Frederick died in 1771, so did the celebrated 'age of liberty'. His successor, Gustav III, took advantage of party political discord and mounted a royal coup against the Diet to reinstate the Crown as the head of the executive government, restricting the Diet to budgetary and legislative functions. The coup was swift and successful, and lack of popular resistance meant that it remained entirely bloodless.

Charles Francis Sheridan, the secretary to the British envoy in Sweden, gave a popular account of the revolution. His *History of the Late Revolution in Sweden* was a reminder of the dangers of despotic rule and the threat it posed to the security of European liberties. In a single day, he reported, the revolution 'converted a government, supposed to be the most free of any in Europe, into an absolute monarchy'. This was the beginning of the end of Sweden's Enlightenment. Sheridan felt that the ability to defeat and overthrow the yoke of despotism was proportional to the degree of enlightenment in a nation. The history of revolutions revealed what qualities were requisite for political and intellectual freedom, and taught

that liberty would flourish in proportion as the minds of men become enlightened: that in an age, in which the principles of society itself, have been considered as a science; the nature of government analyzed, ascertained, and reduced in some measure to a system: when, consequently, in proportion to the progress made in this science, and to the general increase and diffusion of political knowledge among mankind, the benefits resulting from freedom, must not only be more universally known, but likewise the means of acquiring or preserving it, better understood.[10]

Enlightenment and liberty were precarious privileges, not – it was increasingly evident in a number of examples from around Europe – powers that inevitably prevail 'universally throughout Europe'.

The political changes effected by the royal coup were disturbing to those who believed that this would lead back to monarchical despotism. Gustav III, however, maintained otherwise and seems to have sympathised with ideals of popular sovereignty and a balance of power between the Crown and the Diet. The preamble to the new constitution of 1772 declared 'the greatest abhorrence to a king's despotic power' and proclaimed the king's commitment to promote the strength, welfare and happiness of the loyal subjects.[11] But the outbreak of the French Revolution in 1789 brought internal opposition against the Swedish king to a head. Fired by the battle cry of the French revolutionaries, young Swedish aristocrats charged themselves the task of regaining their constitutional rights to rule, which they felt were wrongly usurped by the monarch. Ironically, in 1790 Gustav shied away from leading an expedition into France which was to be an attempt to restore the Bourbons to the throne, purportedly claiming, 'I am myself a democrat.'[12] Yet, disaffected young noblemen did not abandon their political grievances against Gustav, and in March 1792 the king was assassinated.

Coup d'état, revolution, regicide: these dynamic events easily captured the intrigue of travellers surveying modern Europe. While accounts of past territorial disputes and border control informed travellers' historical consciousness, the past paled compared to the colourful affairs of the present. The outbreak of the French Revolution had an important effect on travellers' interests in the Scandinavian countries, but not merely because France – the traditional port of entry to Europe for the British travelling abroad – was off limits, thus forcing a northern route. Sweden was a destination itself, with a recent political history that made its contemporary government structure of interest for comparative analysis. Might Sweden suggest an alternative model to how Enlightenment principles could be used to improve society? How did the Swedish government cope with their 'revolution' of 1772, the redistribution of political power and the assassination of the king? How was Sweden shaping up in the wake of the 'age of liberty'? In the aftermath of the French Revolution and the shocking rise of Napoleon's regime, seeking answers to such questions was well worth the northern tour.

But, of course, the French wars did more than spark interest in, and promote political awareness of, the northern governments. Particularly after hostilities broke out between Britain and France in 1793, the route to the Baltic was vital to British communications with allies and to European trade.[13] But the British also grew familiar with northern natural resources, industry, culture, customs, art and architecture.

A land of wood and iron

Like most major European cities, Stockholm was not unfamiliar to foreign dignitaries and travellers, whose letters and reports home represented quite prominently the cultural and political affairs of Sweden's capital. With the exception of Uppsala, the quaint academic town about 100 miles north of Stockholm, the rest of Sweden was absent in travellers' accounts and remained largely unexplored territory. Indeed, this was true for most of Norway and Finland as well, so that Sir John Carr (a lawyer-turned–popular travel writer), who toured round the Baltic in 1804, wrote:

> The ground which my pen is about to retrace, has not very frequently been trodden by Englishmen. Northern travellers of celebrity, who have favoured the world with the fruits of their researches, have generally applied their learning and ingenuity more to illustrate the histories of the countries through which they have passed, than to delineate their national characteristics.[14]

The move from the study of history to customs helps distinguish new concerns for travellers and new subjects for publication. Included here were even travellers who were concerned with mapping the meridian, such as Maupertuis and other Arctic explorers, whose accounts dealt more with physical geography than ethnographic information.

Besides reports of scientific expeditions and journeys of discovery, though, another genre of eighteenth-century travel accounts developed, written with conscious efforts to bring together knowledge of natural history with knowledge of the social structure of different countries. In the early nineteenth century, Robert Walpole, who compiled and edited two volumes of articles from various travellers, reaffirmed that the interesting observations were those which 'repay [the traveller] for the difficulties [to] which he is exposed', mainly 'the comparison of the antient and modern geography; mineralogical, botanical, zoological pursuits; the examination of the remains of antient art; observations on the manners and customs of the mixed population of the provinces which he visits . . .'.[15] Clarke's travel narratives, part of which were included with the compilation, fit the bill perfectly.

Clarke wrote in his published narrative that, after a decade's travel, he had 'looked forward with eagerness towards the pleasure he should experience, in comparing the manners of the *Northern* nations with those of the inhabitants of the *South* of Europe'.[16] At the rushed pace at which they were travelling, before they knew it they had run through Hamburg, Copenhagen and Gothenburg, and found themselves in Stockholm (see Plate 1). In the metropolis they shopped for books and supplies, picking up Johann Winckelmann's 'valuable work' *Histoire de l'Art chez les Anciens*, and were proud to have stumbled upon '*Hermelin's* splendid Maps of *Sweden*, and put them into a tin roll for our journey'.[17] The principal feature of their brief stay in the capital was a visit to the arsenal, where they saw a wax figure of Gustav III. The effigy of the 'very handsome man' was preserved in a glass case, near the blood-stained masquerade costume he was wearing at the time of his assassination.

The holes made in the sash and jacket, when he was shot, shew that he was dreadfully wounded in the loins, just above the hip. There is one large hole, through which the principal contents of the pistol were discharged, surrounded by other smaller holes, as if caused by common shot. Even the napkins and rags which were hastily collected at the time of his assassination, to apply to his wound, are here carefully preserved. They exhibited to us the *nails*, the *knife*, and other articles taken from the King's body; also the *pistol* from which they were discharged.

In this morbid wardrobe were the remains of the first act of a drama that commenced the night of the masquerade. The true circumstances of the plot to kill the king were just beginning to unfold, and 'some future *Shakespeare* may find, in the mysterious circumstances connected with the death of *Gustavus*, a plot not unlike that of the Tragedy of *Hamlet*; for which we have been already indebted to the annals and characteristic manners of *Northern* nations.'[18] After leaving the arsenal, they visited the Senate House. Opposite this building was where Ankarström, the assassin, was taken onto the scaffold, publicly flogged and left chained to a post under a sign labelling him the 'Assassin of the King' (inscribed 'in the Swedish language', Clarke pointed out). Clarke obtained one of the portraits that was being sold of Ankarström in captivity (see Plate 2). After three days the assassin was decapitated and quartered, the different parts of his body being displayed around the city on wheels, such as travellers might also encounter while riding through the countryside (see Plate 3).

There was, no doubt, more to tell about Stockholm than stories of revolution, assassination and punishment, but Clarke was also eager to journey to the heart of the north. He desired to wander through the woods and acquaint himself with rural life, and to collect artefacts which would help illuminate the present condition of the environment and its inhabitants. Thankfully for Clarke, Cripps shared his enthusiasm. However difficult it was to travel over foreign territory, with little sleep and in uncomfortable wagons over roughly paved paths, Clarke was impressed with his student's keen commitment. 'Our expedition has succeeded beyond our most sanguine expectations,' Clarke wrote to a friend back at Jesus College, a month and a half out:

> Cripps makes an excellent traveller. He is occupied in sending a case of minerals to Stockholm. Mineralogy, botany, manners, politics, astronomy, antiquities, have all found a place in his journal; he seems to grasp at universal science; and works with his hammer among the rocks, like a galley-slave.[19]

Along with meticulous observations packed into the pages of private journals, letters and, later, published narratives, Clarke and Cripps avidly, perhaps even egregiously, collected artefacts throughout their journey. Sweden stocked a plethora of natural history specimens, which came as no surprise to anyone familiar with the prevalent image of the nation covered with forests. It seemed the essence of Scandinavia. Sweden was known to be a large country, at least three times as large as Great Britain. Its industry was famous; for centuries it was the world's greatest producer of copper and exporter of bar-iron. Close to three-quarters of the Swedish landscape (including Finland during this time) was covered with coniferous softwoods, its greatest natural resource.[20] The forests were effective sites for scientific and technological research, where eighteenth-century Swedish natural philosophers searched for efficient ways to sustain the industrial economy and domestic demands. They made unique innovations in environmental experimentation and engineering, from chemical analysis of minerals to the development of hydrodynamic technologies applied to mining and iron manufacturing.[21] Travelling through the dense, silent expanses of her pine forests became a characteristic image of experiencing the north.

Clarke compared his observations with other published descriptions, particularly from the Swedish botanist Carl Linnaeus's *Flora*

Lapponica, the Norwegian naturalist Eric Pontoppidan's *Natural History of Norway*, and the Swedish mineralogist Gustav von Engeström's *Guide aux Mines*, all purchased especially to be taken on the journey.[22] Linnaeus's and Engeström's texts were essential for any travelling naturalist in Scandinavia, the authors being two of Sweden's most eminent natural philosophers. Pontoppidan's study was held with less regard. Clarke thought the *History of Norway* 'a very jejune performance, and unfortunately the only one that has been translated into English'. Latin and French were standard scholastic languages, and often preferred over many careless English translations of foreign works. But the translation of Pontoppidan's was not the problem. His descriptions based on folklore and myth, his beliefs in – and less than scholarly accounts of – sea-serpents, mermaids and other monsters, however, were taken to be more of a problem.

By all accounts, travellers to the Scandinavian environs outside the metropolises were impressed with the sheer expanse of forested land. Nature was here ubiquitous: *the* defining characteristic of the north. It had an overwhelming effect on people. Clarke jested:

> If the Sovereigns of *Europe* were to be designated by some title characteristic of the nature of their dominions, we might call the *Swedish* monarch, *Lord of the Woods*; because, in surveying his territories, he might travel over a great part of his kingdom from sun-rise until sun-set, and find no other subjects that the trees of his forests. The *population* is everywhere small, because the whole country is covered with wood.... The only region with which *Sweden* can properly be compared, is *North America*; a land of *wood* and *iron*, with very few inhabitants, 'and out of whose hills thou mayest dig brass.'[23]

There was some substance to his satire. As we will see below in the discussions and descriptions of Lapland, theories of what effect the natural environment might have on the human condition were entertained in many medical discussions (see conclusion to chapter 5). But one did not have to be a physician to sense the effects such an environment would have on visitors.

It was vast, cold and lonely. But rather than be rejected as harsh or hostile, the simplicity and organic appeal of the environment inspired many Romantic sentiments. Mary Wollstonecraft's *Letters written during a short residence in Sweden, Norway, and Denmark*, her autobiographical memoir-cum-travelogue (published in 1796), helped popularise the latter impression. Away from her estranged lover Gilbert Imlay, the 36-year-old Wollstonecraft travelled far from home, alone, through exotic fields. Wandering through Scandinavia, she acutely described the environment, wrote of waterfalls, castles and of the comforts and pitfalls of her journey; she recreated portraits of the nobles, peasants and beasts she encountered. Readers with a craze for narratives of travel through the exotic and passions for literary romance consumed her *Letters*.[24] She brought English hearts closer to Scandinavia. As the poet Robert Southey expressed it, her book 'made me in love with a cold climate, and frost and snow, with a northern moonlight'.[25]

Wollstonecraft propagated a view of Scandinavia that made its climate and natural geography themselves therapeutic to those on intellectual journeys. Her readers were asked to appreciate the values not only of exploration, but of nature itself. It was an appropriate sentiment; after all, the study of Scandinavian natural resources gained Swedish natural philosophers international renown. She couldn't avoid remarking that Sweden was 'the country in the world most proper to form the botanist and natural historian'.[26] When Clarke travelled in Wollstonecraft's footsteps four years later, he agreed: 'Sweden would make any man a botanist.'[27]

Travellers might have also jested that visiting the Holy Land would make everyone devout, or visitors to Greece would become marble sculptors. The comments about Sweden were fitting not merely because Scandinavian flora was ubiquitous, but that Carl von Linnaeus, the 'father' of Swedish botany, revolutionised that branch of natural history in the mid-eighteenth century. Linnaeus, professor of botany at Uppsala University from 1741 until his death in 1778, was famous for developing a botanical classification scheme, which he described in his *Systema naturae* (*System of Nature*) (1735). His 'sexual system', so called because the classification of the plant was determined by its sexual organs (the number and arrangement of the stamens and pistils), was simple and practical. In 1753 he published *Species Plantarum*, where he introduced binomial nomenclature – the method of assigning two names to its classification (a general and a specific name), a convention still in use today. (Linnaeus's work was officially recognised in 1905 as the 'starting point' for modern botanical nomenclature.)[28] His botanical work was pioneering and his reputation flourished.

Science and education

Uppsala University's reputation also flourished. As professor, Linnaeus was an inspirational teacher for many students who accompanied him on his celebrated 'herbations', or botanical excursions. He redesigned and carefully cultivated Uppsala's botanical gardens and planted the seeds for the splendour of the gardens that tourists can see even today. He preached agricultural improvement, encouraged manufacturing and asserted the economic utility of science for the state; he was a dominating figure in the 'Golden Age' of Swedish science which mark the most prosperous years of the Age of Liberty.[29]

Linnaeus's reputation was by no means confined to Sweden; and after his death travellers honed in on the places where the famous botanist once lived and worked. William Coxe, who was accompanied by Linnaeus's son around the Uppsala botanical gardens, lauded his achievements. 'The name of Linnaeus may be classed amongst those of Newton, Boyle, Locke, Haller, Euler, and other great philosophers, who were friends to religion: he always testified in his conversations, writings, and actions, the highest reverence for the Supreme Being.'[30] Ranking the Swede amongst the most celebrated (and most pious) English natural philosophers was more than a compliment, it was a gesture towards equalising national prestige. The late seventeenth and early eighteenth centuries were Britain's shining moment, the era of the philosophers who consummated the 'Scientific Revolution'. It was the period when experimental and mechanical philosophy reigned supreme; Newtonianism triumphed over antiquated Scholastic Aristotelianism.

Sweden's eighteenth-century achievements did for the natural sciences what had been done one hundred years earlier for the physical sciences. This was the epoch of what has been termed the 'second scientific revolution'. In the 'age of revolutions' which has come to characterise the late eighteenth and early nineteenth centuries, Sweden's scientific achievements can be thought to mark a point of origin of *modern* science.[31] Linnaeus's success in propagating his new classification scheme was complemented by the achievements of his compatriots. One could here think of the philologist Eric Benzelius; Anders Celsius, who invented the centigrade thermometer (and joined the French expedition to the Arctic Circle, led by Maupertuis, to prove Newton's theory of the flattening of the poles); the philosopher, religious leader and spiritual explorer

Emanuel Swedenborg; and the advances in laboratory science and the chemical analysis of minerals by Johan Wallerius, Axel Fredrik Cronstedt and Torbern Bergman. For science, the northern lights shone bright in eighteenth-century Sweden, and they attracted widespread European interest and respect.

In November 1799, Clarke and Cripps, with their horses and carts loaded with trunks and filled with 'a variety of minerals, plants, insects, antiquities, Drawings, Maps, &c.', along with an entourage of servants, bodyguards and translators, approached Uppsala. 'A long avenue of stately firs at length opened upon Upsalia, once the metropolis of all Sweden', Clarke recorded.

> Its appearance, in approach to it, is really noble: we descended a hill towards it, calling to mind the names Celsius, Linnaeus, Wallerius, Cronstedt, Bergmann, Hasselquist, Fabricius, Zoega, and a long list of their disciples and successors, which has contributed to render this University illustrious; the many enterprising young travellers it has sent forth to almost every region of the earth; the discoveries they have made, and the works of which they are the authors. For since the days of Aristotle and of Theophrastus, the light of Natural History has become dim, until it beamed, like a star, from the North; and this was the point of its emanation.[32]

They remained in Uppsala for about one week, absorbed in the glory of the past. The shining prestige of Swedish science in its 'Golden Age' was reflected in their thoughts.

The foundations of institutions in the eighteenth century, such as the Swedish Academy of Sciences (f. 1739) facilitated contacts and the exchange of ideas and technologies with similar institutions abroad. Swedish scholars were elected to foreign academies. Benzelius, Linnaeus, the South Seas traveller (and Linnaean 'disciple') Daniel Solander and Bergman were among those elected foreign members to the Royal Society in London. The Swedish East India Company, referred to by the English counterpart as 'by far the best regulated and prosperous in Europe', had made salient strides in exploration and exploitation of foreign natural resources.[33] The 'north' was not peripheral but a nodal point in the network of intellectual communications. As one historian of Sweden's Age of Liberty summed up: 'No doubt Swedish upper classes assimilated, more readily and more generally than ever before, European intellectual and cultural influences, European civility.'[34] Under the patronage of Gustav III, a valiant attempt was made to enhance even further its national prestige. Systems of patronage overall were of interest to British travellers in Sweden. In 1804, while gazing admirably at the art and architecture which charmed Stockholm, John Carr happily observed that:

> most of the living artists of Sweden owe their elevation and consequent fame to the protective hand of the late king, Gustav III, a prince, who, to the energies and capacities of an illustrious warrior, united all the refined elegance of the most refined gentleman; . . . What Frederic the Great was to Berlin, Gustavus the Third was to Stockholm: almost every object which embellishes this beautiful city arose from his patronage.[35]

Had he been thinking of Uppsala when he applauded the king's patronage, he no doubt would have recognised that science was also a beneficiary of the king's generosity. Inspired by patriotic zeal, the Court eagerly pursued the fashioning of polite society in all different manners. Coxe was witness to the king's court ceremonial:

> The strictest adherence to form is observed in this court: while many sovereigns of Europe are endeavouring to retrench the ceremonies attendant on royalty, Gustavus III has introduced a degree of pomp and etiquette familiar to that used at Versailles, and hitherto unknown in this country. The king appears to possess too large an understanding to be, in this instance, a servile imitator of the French; it is therefore more probable that his motive for this conduct is in some measure political, and the increase in royal prerogative may have rendered it expedient to throw an additional splendour round the majesty of the throne.[36]

It was perceptive to compare the Swedish court with Versailles – a clear model for Gustav. But the opulence was not for vanity. At the time of his father's death in 1771, the then Prince Gustav was in Paris, hobnobbing with the *philosophes* and entertaining himself at the literary *salons*, where the atmosphere buzzed in intellectual discourse. After his coup, the new king refashioned himself after Henry Bolingbroke's *Idea of a Patriot King* (1749), which called for a monarch powerful enough to end party divisions and revitalise national liberties. Gustav's commitment to return Sweden to the kind of political glory it had experienced one hundred years earlier was intended to be facilitated through the promotion of the arts and sciences. He wanted artists and philosophers in Sweden to imbibe an intellectual atmosphere. Perhaps the failure of England's George III to fulfil Bolingbroke's vision of a 'Patriot King' after his accession to the throne in 1760 was one reason why English literary travellers and historians ruminated over Gustav's reign.

When Clarke travelled through Sweden, he could see only the remnants of Gustav's forays into the public purse. The king's munificence – however well intended – were not altogether justified considering the country's lack of fortune. Yet one endowment to Uppsala University Library from Gustav particularly captured Clarke's curiosity.

> It consists of two chests of manuscripts, double-locked, chained, and sealed, which are not to be opened until fifty years shall have elapsed from the time of his death. These chests are supposed to contain his foreign correspondence, and many papers relating to the principal transactions in which he was engaged and the state of *Europe* at the time of his reign. An *English* traveller will hardly participate the feelings of curiosity which are betrayed by the *Swedes* respecting these mysterious boxes. 'What a misfortune for us,' said one of the inhabitants of Upsala [*sic*], 'that this precious deposit will not be opened in our time.' Great expectation is on foot with regard to the things that will come to light when these papers are examined.[37]

It was unfortunate that this literary traveller visited Sweden after only seven years had elapsed. The king's foreign correspondence and documents relating to the state of Europe looked indeed be a treasure chest for the surveyor of modern political relations.[38]

That particularly intriguing historical artefact may have been inaccessible, but the annals of the University had plenty of other memorials for travellers to explore. When Coxe had travelled through Uppsala, Torbern Bergman, the professor of chemistry, showed him around the famous chemical laboratories. Bergman was something of a polymath, 'whose fame stands among the learned of all nations, and whose reputation is deservedly established for his useful and accurate researches, as well in his own laboratory as in every branch of natural history'. He had demonstrated skill in botany (even capturing a gold medal over Linnaeus for an Academy of Sciences competition concerning worms on fruit trees) and geology, but his reputation was rooted in his chemical researches. He reformed chemical nomenclature, developed a theory of chemical affinity and chemical combination, and demonstrated the existence of carbonic acid, which led to his preparation of mineral water (which was soon after commercially produced in other European countries).

By the time Clarke arrived in Uppsala, though, Bergman was dead. Instead, he was taken around the now enshrined laboratories by the new professor of chemistry, John Afzelius. An added, and impromptu, bonus was joining the students to attend some of Afzelius's public lectures. Clarke was impressed.

> Around the chemical lecture-room were arranged the Professor's collection of minerals, – perhaps more worthy of notice than anything else in Upsala; . . . It was classed according to the methodical distribution of Cronstedt, and has been in the possession of the University ever since the middle of the eighteenth century. The celebrated Bergmann added considerably to this collection, which may be considered as one of the most complete in Europe; especially in specimens from the Swedish mines, which have long produced the most remarkable minerals in the world. . . . One small cabinet contained models of mining apparatus; pumps, furnaces, &c. there is no country that has afforded better proofs of the importance of mineralogical studies to the welfare of a nation, than Sweden.[39]

Encountering a mineral collection based on Cronstedt's methods, who, as Clarke went on to note, 'laid the true foundation of the science, by making chemical composition of minerals the foundation of the species into which they are divided', was commensurate to strolling through Linnaeus's gardens and seeing how the botanical kingdom was so masterfully and systematically ordered. The delight expressed by the guests when proudly shown the scientific collections was more than an acknowledgement of the intellectual labour deployed to name and classify nature's specimens. Visitors showed esteem for the extent of the collections themselves. Through the unflagging efforts of her miners and natural historians, Sweden's rarest gems and natural wonders were being discovered, analysed and used to exhibit and publicise the natural wealth of the nation. These were collections and researches clearly fit for national pride.

But however exemplary the collections, the frontiers of scientific knowledge were located beyond the walls of the ivory towers. Clarke was anxious to travel to the field and visit sites of active exploration and excavation. This is where he particularly differed from previous British travellers, such as Coxe. More concerned to evaluate the present condition of society than revisit ancient history, Clarke preferred mines to crypts, classification schemes to genealogical charts. Thus with Engeström's *Guide aux Mines* in hand, Clarke and Cripps journeyed north-west from Uppsala to the Great Copper Mine at Falun, the largest and most productive mine in Sweden, which also boasted a technical college, the *Laboratorium mechanicum* (see Plate 4). There they met Johan Gottlieb Gahn, head of the research school and who, since 1784, was assessor to the Swedish Board of Mines.[40] Gahn was occupied in extending experimental researches in the chemical composition of minerals, using techniques pioneered by Cronstedt, his mentor. Gahn was the acknowledged leading practical chemist of his day, and had collaborated extensively with Carl Scheele (one of the first people to identify oxygen as a chemical element) and Bergman concerning new experimental techniques for the chemical analysis of minerals. Gahn also held one of the most highly regarded pedagogical posts in late eighteenth-century Sweden, being responsible for training civil servants in proper methods of chemical analysis for their onsite work in the mines.

During his few weeks in Falun, Clarke actively participated in the training regimes and learned cutting-edge experimental techniques. Gahn tutored Clarke in chemical analysis with a 'blowpipe', the forerunner of our modern-day Bunsen burner. Blowpipe analysis was pioneered in Sweden, and was instrumental in effecting the 'chemical revolution in mineralogy'.[41] As Clarke found, the instrument looked seductively simple: it was merely a small, hand-held metal tube. But to use it correctly required a great deal of skill. By carefully blowing through the pipe, the practitioner could direct concentrated air into a candle flame, which was in turn directed onto a small mineral sample. If the mineral melted, information could be gleaned about its chemical constituents.[42] Blowpipe analysis became crucial to the mining industry in Sweden, where it was used to identify the metallic content of ores which could lead to more efficient mining practices and higher profits. Because of its portability, individual-oriented management, economy of use and quick results, Swedish miners were equipped with the instrument and carefully instructed in its use. It was also these techniques, however, that Bergman, Scheele, and Gahn applied to the development of their mineral classification systems. Clarke was receptive to the advantages offered to travelling naturalists by this instrument, and acquired one which he used to analyse the landscape throughout the rest of his travels. This marked the beginning of his mineralogical education that would eventually culminate with his obtaining a Cambridge University professorship in mineralogy in 1808. This also marked a significant development in the ways that Clarke advocated how one should go about travelling: scientific analysis now became part of the way he thought that one should acquire information about foreign lands. The subsequent results of his mineralogical analyses were recorded in chapters of his Travels. His mineralogical survey of Scandinavia would not be matched by another British traveller until just over ten years later, when Thomas Thomson, later to become professor of chemistry at University of Glasgow, went to see what advances in science had been made there since the time of Clarke's trip.[43]

But Clarke was pessimistic about the future of Swedish science. He judged the state of science as manifested in the activities from Stockholm to the universities to be in decline, since much of their patronage had been lost after Gustav III's assassination. Uppsala's eighteenth-century naturalists pioneered an understanding of their 'land of wood and iron . . . because *natural history* is almost the only study to which the visible objects of such a region can be referred'.[44] But the names of its illustrious alumni recited by Clarke with such blissful admiration were products of a bygone era. Science had flourished under a supportive regime. Linnaeus's 'disciples' were well trained and sent to almost every region of the earth, from Japan to South America, to explore and collect specimens. Linnaeus promoted a special form of 'scientific travel', requiring his protégés to proselytise others to his system of classification. He argued that Sweden's success as a scientific nation would be secured only through comprehensive comparative study of nature around the world: scientific travel was patriotic travel.[45] This, Clarke told his readers, England should learn from.

But despite the commendatory 'enterprising travellers' that had rendered Uppsala University illustrious, the northern light was once again dim. He suggested that travellers of the nineteenth century needed more intellectual stock to amalgamate a cosmopolitan view of natural history, politics and culture. This was certainly what defined an educated *English* literary traveller. Clarke sized up the scene by commenting that 'Centuries may elapse before *Sweden* will produce a *Locke*, or a *Montesquieu*, or a *Paley*, or a *Dugald Stewart*; although it may never be without a *Wallerius*, a *Hasselquist*, a *Thunberg*, or a *Berzelius*.'[46] The duty of the *modern* traveller, such as Clarke, who had at Cambridge studied Locke and Paley, and was certainly familiar with the French and Scottish social theorists, was to take from Sweden, and everywhere else visited, what could be appropriated, intellectually and materially.

Clarke's comments about the decline of Swedish science, which he linked to declining patronage, were woven into a wider argument he made about the future of English science. He hoped Cambridge would one day become what Uppsala had been, a scientific 'metropolis', the centre from which the rays of knowledge of all luminaries of natural philosophy should emanate. For this to happen, science education must be seen to bolster national prestige, and it must receive state patronage. After all, that appeared to be the key to Sweden's success earlier in the eighteenth century.

Clarke's polemic for the advancement of science education was written with pomp and authority. Already successful at lecturing on chemistry and mineralogy by the 1810s (when writing about this in his *Travels*), he was further able to reflect on the foreign institutions he had visited and invoke his experiences in comparing different institutional structures to add weight to his argument. As in Sweden, England should better support the researches of her natural philosophers. The benefits, he suggested, were not only that the character of its Universities would improve, but that it would ultimately lead to economic advantage.

He illustrated his points with reference to his visits to Swedish scientific institutions. At the Public Seminary at the Kongsberg mines, for example, three resident professors provided instruction in mineralogy, geology, chemistry, physic, mathematics and other branches of science.

> Any of the miners, or children of the miners, may attend this institution. . . . FOR THESE LECTURES, NO PAYMENT WHAT-SOEVER IS REQUIRED. . . .We felt, at that moment, an inward sense of shame for our own country, in which such studies have hitherto met with so little encouragement. We would but turn our thoughts homeward, and ask, what the Government of GREAT BRITAIN had ever done towards the advancement of *mineralogical* knowledge.[47]

Clarke, sensing the irony of the passage since he was now a professor of mineralogy, conceded in a footnote to that passage that over
the decade or so since he visited Sweden, science education in England
had more favourable prospects. Of course, there were stark differences between Swedish and modern English educational programmes
that Clarke felt needed to be drawn out:

> When an *Englishman* speaks of the Universities of *Sweden*, or when
> he is reading the different accounts that have been published of
> *Upsala*, it is not often that any right notions are entertained,
> either of the Seminary that bears this name, or of the habits
> and tact of the Students and Professors. If, for example, he forms
> his notion of a *Swedish* University from any thing he has seen
> of similar establishments in his own country, associating ideas
> of *Cambridge* and *Oxford* with his imaginary conceptions of *Upsala*,
> *Lund*, and *Åbo*, he will be egregiously in error. It is not easy to
> conceive any thing more foreign to all our notions of the dig-
> nity and splendour of a national seminary for education, than
> in the real state of things in *Upsala*. Perhaps there may be some-
> thing to compare with it in the Universities of *Scotland*; but even
> in the last there is nothing so low as in *Sweden*.[48]

This was a rather critical, and somewhat vainglorious, comparison
of education in different nations. Clarke here reconfirmed the intellectual distance between England and Scandinavia. But the contrasts
between the various institutions were to provide not merely another
example of the alleged decline of Swedish education, but of the
opportunities to support scientific education at different levels. Clarke
lamented that 'field' academies, institutions similar to those at
Kongsberg and Falun which offered instruction in the sciences and
experimental practices to labourers, were not established in England.
These, however, were distinct from the education received at universities such as Cambridge, where students – the mine *owners*, the
landed gentry – should cultivate knowledge of the principles of
science. Education for 'national improvement', gained from agricultural, industrial and moral improvement, should be centralised
in a university setting: these educational ideals, part of the philosophy of 'Enlightenment improvement', were central to convincing
the social élite that they were pursuing the right road to progress
and prosperity.[49]

Political economists, including Clarke's erstwhile travelling companion, Thomas Robert Malthus, likewise pressed principles of progress and the concerns over scientific improvement, particularly agricultural, into educational ideals. On one level, Clarke's discouraging view of the current state of Swedish intellectual affairs seem to be corroborated by Malthus. Malthus's views, however, were more complicated, and his opinion of current Scandinavian affairs varied as he traversed through different countries of the Northern Kingdom.

Population, production and progress

After one month of travelling together, and just after entering Sweden from Denmark, Malthus and Otter took their leave of their travelling companions, Clarke and Cripps (who travelled toward northern Sweden), and headed for Norway. Both parties pursued a vigorous tour of Scandinavia, occasionally crossing paths.

This was Malthus's first trip abroad, during which he was collecting statistical information in preparation for the second edition of his controversial *Essay on the Principle of Population*. Plans to arrange a Scandinavian tour were particularly fitting for Clarke and Malthus since both had desires to gather information about the current affairs of the northern states. But the two shared more than their interests in fact-finding. Both were from Jesus College, Cambridge, where Malthus graduated 9th Wrangler (ranked ninth in the list of graduates with mathematics honours) in 1788, which was also when he took Orders for the Church of England. In 1793 he was appointed to a college fellowship. Through their college connections Malthus and Clarke remained in contact long after their trip to Scandinavia. Upon their deaths, their mutual friend and travelling companion, William Otter, was the first biographer of both.

Like Clarke, Malthus was an avid reader of travel narratives about the northern countries, was aware of their history and concerned about the balance of power in the states surrounding Napoleonic France. The topics of inquiry that have been introduced in this chapter so far – revolution, improvements to society, the making of the 'modern' – loomed large in these travellers' minds. Because Malthus's work became highly influential to nineteenth-century debates about government rule, population and social welfare, and since he co-ordinated his journey with Clarke, it is worthwhile paying some attention here to how his Scandinavian travels informed his political philosophy. This will also provide a way to contrast different opinions of various travellers regarding Swedish government patronage of science and medicine, and the different ways that the experience of travel was used to support their theoretical assumptions of what defined modern states.

Malthus's interests in Scandinavia concerned, in broadest terms, the 'health' of the economy, agrarian production and population reproduction. The health of the economy affected the health of the population, and the political management of the land affected both. As befitted the philosophies of national improvement and political progress, economic questions became integral to broader concerns about the condition of civil society. The study of comparative national economy flourished in the later eighteenth and early nineteenth centuries in the work of economists and philosophers such as Adam Smith, David Hume and David Ricardo, and agricultural 'improvers' including Arthur Young, William Marshall and Count Rumford (Benjamin Thompson). Theirs was a study in 'political economy' – an examination into the political management of natural resources which underlie a nation's wealth. The links made in political economy between state management, individual labour and responsibility, and the moral and material prosperity of society made it another appendage to the 'sciences of man'.[50] Many different subjects informed discussions in political economy. War, statistics and demography, medicine, public health programmes, agrarian improvement and evangelical philanthropy were sum and substance of Enlightened inquiry. Malthus's work was instrumental in preparing the ground for other studies central to population questions, social improvement and political economy in the nineteenth century.

It is generally recognised that from the early eighteenth century more demographic data were being widely collected and used to compute prices of annuities, life insurance and population statistics.[51] But in 1749, Sweden in particular embraced the 'quantifying spirit' by instituting an official policy requiring local pastors to collect, at five-year intervals, data on birth and mortality rates, along with marriage statistics. The government used these records to argue the need for an increase in population. An optimistic plan that was promoted by mercantilist philosophy proposed that through population growth and public health measures, the wealth and power of the state would grow.

Mercantilist optimism influenced another philosophy of human progress and perfectibility espoused by the so-called 'utopians'. The eighteenth-century utopian philosophers believed that in a perfectly organised society, a population of any size would be able to share its resources and avoid imposing government controls or policing of property for the equal distribution of subsistence. After the outbreak of the French Revolution, questions of perfectibility of human kind, equality, government control over natural resources, and so on, gained political potency, and an array of 'scientific' judgements about the history and progress of civilisation was deployed to analyse contemporary social conditions. In this context, the Marquis de Condorcet in France espoused the utopian philosophy. During the Reign of Terror in 1794, Condorcet, under proscription from Robespierre and in hiding, wrote the work for which he is best known, the *Esquisse d'un tableau historique des progrès de l'esprit humain* (1795; *Sketch for an Historical Picture of the Progress of the Human Mind*). In it, he proposed that the human race had gone through great 'epochs' of history, in which humans have been in continuous progress from a state of savagery through levels of increasing enlightenment and happiness. In a future epoch, he argued, inequality between nations, classes, and ultimately individuals would be eliminated, and humans would enjoy intellectual, moral and physical perfection.[52]

The English social philosopher and political journalist William Godwin had expressed a similarly optimistic view of the future perfectibility of humankind. In his *An Enquiry Concerning Political Justice, and Its Influence on General Virtue and Happiness* (1793), he advanced a doctrine of extreme individualism, arguing that small self-subsisting groups should replace large government and social institutions to ensure the future improvement of society. In his radical philosophy, he argued that all forms of social organisation (judicial, religious, educational, and so on) were oppressive, and after their removal humankind would be liberated from misery, ignorance and poverty, and live in a world of uncoerced morality, virtue and happiness. Godwin anticipated a potential population problem, and predicted that humans would 'probably cease to propagate' and 'perhaps be immortal' after their liberated rationality eliminated sexual passions.

Malthus's view was quite different. Where Condorcet and Godwin entertained visions of progress, peace and plenitude, Malthus focused on population problems, poverty and public policy. He believed that population growth would inevitably exceed levels of subsistence necessary for survival. His views were largely compatible with the philosophy of the 'physiocrats', who believed that a nation's wealth was attributable to agricultural productivity, and that population growth itself would not lead to increased wealth. The physiocrats objected to pronatalism, arguing that human reproduction should not be encouraged to a point beyond that sustainable without widespread poverty. They also objected to any other strategies involving the intervention of the government to promote population growth, leading to the pronouncement of their famous 'laissez-faire' policy which became so influential to the thinking of later classical economists. After reading Condorcet and Godwin in 1797, Malthus concluded that the notion of a perfectible society was a theory built on sand.

Malthus wrote the first edition of his *Essay* as a critique of the utopian theories of Condorcet and Godwin, and criticised what he considered their unsound principles of reasoning. He attacked Condorcet's theory of the 'indefinite' perfectibility of humankind for being entirely speculative, ambiguous and devoid of proper methods of investigation and authenticated proofs. The future epoch envisaged by Condorcet might have been a noble dream, but it was unrealistic to suppose that the capacity for biological improvement had no limits.[53]

Similarly, he criticised Godwin's fanciful notions of human 'immortality', the proscription of all emotions and passions, and the pervasive control of the mechanisms of rationality, which Malthus dismissed as figments of the imagination. But the question remained: how could society balance growing numbers with available resources to remedy perpetual cravings? Godwin's thought regarding the increase in human population, which he conceded could take place for 'myriads of centuries' before perfectibility, was merely that 'There is a principle in human society, by which population is perpetually kept down to the level of the means of subsistence.' But what was that principle? Malthus's *Essay* of 1798 was his forthright attempt to answer this question.[54]

The first essay was written as a polemic. It contained the premises of his theory about the balance of population and subsistence, but presented little supporting evidence. In general language, he discussed the fluctuations in populations of ancient society, and amongst primitive peoples, tribal communities and nations of shepherds, Indians, Hottentots and hunters. But he was also concerned about the happiness and health of modern populations, and shared the Enlightenment view that social improvement and progress needed rational management. He argued that moral improvement and civic responsibility were ways to overcome the problem of there not being enough food to go around. As he put it: 'Evil exists in the world, not to create despair, but activity.'[55]

Reviewers of his essay were outraged at Malthus's suggestion that God invited too many to nature's feast, and many dismissed his philosophy as pompous and impious; worse still it was 'a specious argument inapplicable to the present state of society'.[56] Malthus was determined, when preparing his second edition, to demonstrate its applicability to modern society. He wanted his 'principle of population' to be accepted as a natural law, with all the universality and constancy in its application which characterises scientific fact. This required empirical evidence and demonstrable proof. He needed first-hand experience of the condition of modern states, and a journey to Scandinavia provided the perfect opportunity to gather population facts. Concerns over population maintenance, government assistance to the poor, notions of progress in society, health of civic and political bodies all bore upon his philosophy and his observations when travelling through Scandinavia.

Population health: Norway and Sweden

Before travelling, Malthus had some preconceived ideas about what he might find. In the first edition of his *Essay*, it had entered discussion in reference to population statistics on 'Sweden, Norway, Russia, and the kingdom of Naples' which were briefly cited from Richard Price's opus on birth and mortality rates (that work being used to recommend premiums for life insurance and annuities). Multiple editions of Price's work were published by the end of the eighteenth century, but knowledge of the 'Swedish tables' (their population statistics) remained scant until Malthus's visit.[57] However, the *availability* of statistical resources was inviting to Malthus, who saw in them an opportunity to hammer out an empirically supported law of population growth. The structure of the second edition of the *Essay* reveals the extent to which his journey to Scandinavia and the compilation of data from other sources assisted his argument.

The new *Essay* was divided into four 'books', all published as part of one quarto volume. The first book examined the checks to population in antiquity and amongst 'savage' peoples, where he reiterated the existence of polygamy in Africa, incessant war in the 'Kossacks', and promiscuity amongst the South Sea islanders. Competition for food resulted in cannibalism, famine, plague, epidemic and endemic maladies ('positive' checks to population). But people in modern Europe had a chance. The 'civilised man hopes to enjoy, the savage expects only to suffer'.[58] However, hopes of happiness would only materialise if proper preventative measures were taken to balance population growth and its subsistence.

Malthus's liberal use of travel literature which discussed distant parts of the world enabled him to pack detail into his principle. But he intended to be most persuasive by using his own travel narratives to analyse the condition of *modern Europe*. He began with the discussion of the northern countries, including Russia in book two of the second edition, a critical place within the significantly revised structure. His travel diaries reveal the extent of his Scandinavian tour – as far north as Trondheim in Norway, down the Norwegian–Swedish border and over to Stockholm – where we see that information on different customs, produce, and population was collected *en route*.[59] After his tour, Malthus polished his prose from his travel diaries, and moved much of the material directly into his new chapters on Scandinavia, giving the *Essay* the flavour of a first-hand, empirically proven, account of the applicability of the principle.

Malthus stated that the powerful tendency for population to increase applied as much to northern Europe as anywhere, but that particular kinds of preventative checks had prevailed to keep the population under control. What were these preventative checks? Norway and Sweden provided two different models, which were useful for comparison.

In Norway, it appeared that the 'mortality is less than in any other country in Europe', 1 to 48, as shown in the Norwegian census taken in 1769. This was also close to the figure given by the former Professor of Statistics in Copenhagen, Frederik Thaarup, whom Malthus met in Norway and discussed 'the causes that had impeded or promoted the population of Norway'.[60]

It appeared to Malthus that the Norwegian government had a particular commitment to promote agricultural labour which reinforced a spirit of survival and led to moral improvement and healthiness. Eric Pontoppidan had described the self-sufficient Norwegian farmers as 'polypragmatic peasants', whose work, which involved tanning, weaving, carpentry, joining, shoemaking and other regular duties, was a largely undifferentiated labour force which retained a spirit of communal responsibility and mutual aid.[61] In his travel diary, Malthus noted the opinion of a Norwegian gentleman that attention to the improvement in agriculture was a primary concern in Norway. 'One of the most powerful reasons of the present prosperity of the country he thought was, that the people now depended less on fishing, & more upon the produce of the earth.'[62]

The self-sufficiency of the peasants and their spirit for improvement were crucial characteristics to Norwegian society. Agricultural resourcefulness was the driving force to a better society. Necessity commanded efficiency; producing food helped keep at bay nature's afflicting visitations. To him, agricultural labourers yielded real value to society, as agricultural improvement was 'the *sole* species by which multitudes can exist'.[63] In a country such as Norway, which concentrated labour in agriculture, the lower classes were provided with a more comfortable lot in life and lower mortality than might be expected in the harsh northern climate. But, in addition to the 'productive spirit' which drove agricultural improvement, there also existed what he termed *preventative* checks to population growth, which were crucial to maintaining a healthy population. These checks involved the rigorous social control exercised by local parishes over marriages, and particular working conditions that affected labourers earning enough to support a family. As a guard against peasants marrying too young, Malthus was informed that the local parish continued to impose regulations over marriage, giving the pastor the power to refuse to marry a couple thought not to be adequately prepared. Custom prevailed to discourage early marriage.[64] Larger farms could employ 'house-men', a married labourer who, in return for lower wages, was given a plot of property sufficient to maintain a family. But these positions were few, and with 'little variety of employment ... [the peasant] must feel the absolute necessity of repressing his inclinations to marriage till some such vacancy offer'.[65]

Combining parish surveillance, the traditional custom of caution over premature marriage and the limited opportunities of employment were preventative checks to population growth. Awareness of personal responsibilities, 'repressing inclinations to marriage', amounted to what Malthus termed 'moral restraint', a phrase introduced in the second edition.[66] To marry without prospects was immoral. The concept was used to argue that moral responsibility, which was linked to marriage restraint and hard work, was an avenue which led to population control without nature's harsh 'positive' checks. His chapter on Norway was offered as a warning about the consequences of early marriages for population explosions, and proof of the efficacy of moral restraint. Malthus was full of praise for Norway's management of population and natural resources. 'Norway is, I believe, almost the only country in Europe where a traveller will hear any apprehensions expressed of a redundant population, and where the danger to the happiness of the lower classes of people from this cause is, in some degree, seen and understood.'[67] It was a country perfectly prepared to accept the dangers of the principle of population, and undertook acceptable measures to ensure they maintained a healthy and happy society.

If Norway was the place for the Malthusian 'utopia', then Sweden represented the Malthusian 'dystopia'. One striking contrast was that employment in agriculture was not so complete in Sweden as it was in Norway. That, he concluded, explained why 'the positive check has operated with more force, or the mortality has been greater' in Sweden.[68] Despite their similar climate, Norway's population was healthier than Sweden's: in Norway, the people wore 'faces of plenty and content', in Sweden they were 'absolutely starving'.[69] This boiled down to fundamental differences in government attitudes and administration over the maintenance of the health of the population. The prevailing defect in Sweden was the 'continual cry of the government for an increase of subjects, [which tended] to press the population too hard, against the limits of subsistence, and consequently to produce diseases which are the necessary effect of poverty and bad nourishment'.[70]

As we saw above, the Swedish government was keen to collate population statistics in order to determine the rate of population growth and health. The concern to collect demographic statistics was rooted in a dominating mercantilist economic philosophy. This emphasised that accelerating population growth would increase the nation's economic wealth and political power. More people created more demand, and more productive labour to exploit natural resources, pay taxes and defend the nation.[71] Justification for mercantilist population theory could, and often did, refer to the biblical injunction to 'be fruitful and multiply'. Most were not concerned with possible adverse effects of population growth, being confident that any level of population would be capable of securing appropriate levels of subsistence.

Malthus disagreed. Like Norway, Sweden was not by default an unhealthy country, but, he argued, it was prone to unfavourable ('positive') checks to population for reasons relating to agricultural investment and their promotion of public health. First, Sweden was too dependent on its own agricultural production, neglecting assistance that could be gained through importation or more concentrated efforts for agricultural improvement. This resulted in particularly 'unhealthy years', when the harvest season was adversely affected by the climate. Here Malthus benefited from the demographic statistics, the 'bookkeeping of life and death', which had been collected since 1749.[72]

In Stockholm, Malthus visited the Royal Swedish Academy of Sciences, met the secretary to the Society who was also secretary to the Board of Statistics, and had the opportunity to see at first-hand the folios which contained the collated data sent from local parishes. When he returned home, he was able to use these unique statistics to demonstrate the correlation between mortality decreases and increases during respective years of good and bad harvests. In the case of Sweden, the data 'clearly proved' that there, as elsewhere, population had a tendency to increase, but that, because Sweden lacked an efficient system of agriculture, there quickly emerged a poorly nourished, redundant and unhealthy population.[73] Thus, the increase in population was solely 'productive of misery'.[74] A fruitful remedy could be the improvement of the political regulations of agricultural production, most profitably leading to better knowledge of crop rotation and of fertilising their lands. Instead, it appeared to Malthus that the Swedish government was pursuing the misguided course of attempting to treat the sick rather than address the cause of sickness.

In a number of reports to the Swedish Parliament, the *Collegium Medicum* (f. 1663; an incipient National Board of Health) argued that a high percentage of deaths could be avoided if the state favoured the propagation of medical officers and institutional health services.[75] In 1752, when the *Collegium Medicum* was first approached by Parliament about the demographic records, the first hospital was opened in Stockholm, and throughout the rest of the eighteenth century over twenty others were established around the country.

Malthus was critical of these developments. He believed that establishing lying-in and foundling hospitals for the care of infants was wasted money. While he acknowledged that in some particular circumstances the establishment of colleges of medicine could prove 'extremely beneficial', he quickly recited his outspoken opinion on the subject: 'Lying-in hospitals, as far as they have an effect, [have a] tendency . . . to encourage vice. Foundling hospitals, whether they attain their professed and immediate object or not, are in every view hurtful to the state.'[76] Elsewhere he was equally austere in his judgement. One of the main reasons for premature mortality, he argued, was that these institutions, 'miscalled philanthropic', eliminated individual responsibility for family health care. Foundling hospitals appeared to present proof that parents were unable to support their families, and, worse still, the hospitals themselves were largely unsuccessful in keeping children alive. His dire conclusion was that establishing hospitals of this kind appeared to be the most effective way to create positive checks to population.

He went on to say that institutions of this kind were unjust to other members of society since, as with other state-supported medical aid or poor relief, they relieved some in society of civil responsibility and concentrated care in a few at the expense of the rest. Malthus's bitter critique of Swedish health care was argued in the same vein as his attack on the English Poor Laws where he was unequivocal: the Poor Laws 'created the poor' in the way hospitals created the sick.[77]

Just like Pitt's Poor Bill in England, socially sanctioned 'philanthropic' institutions threatened the mainsprings of population by encouraging vice and turning a blind eye to moral responsibilities.[78] The best prospects for the future were in committed labour that sought to increase means of subsistence. As Malthus concluded, if the government endeavoured to 'encourage and direct the industry of the farmers, and circulate the best information on agricultural subjects, it would do much more for the population of the country than by the establishment of five hundred foundling hospitals'.[79]

Many contemporary reviewers of Malthus's second *Essay* rejected his conclusions about the benefits of state-supported health programmes and the benefits of medicine to population health. These debates fall beyond the scope of this study, but it is interesting to note that few seem to have criticised his use of empirical 'facts' about the rates of population grown and his observations about checks to population abroad.[80] His Scandinavian tour was, on the one hand, designed as a fact-finding mission to help him overcome his critics' objections that the first *Essay* was too theoretical and 'inapplicable to the present state of society'. His observations, data and personal interviews were woven into the narrative of the second edition to create a compelling argument that made his revised argument resemble a travelogue more than a theoretical treatise. The condition of Europe was viewed relative to the effects of his 'law of human nature'. His comparisons of health, work regimes, climate, agrarian cultivation and government administration within Scandinavia, and between the northern with other European countries, helped him hone his illustrations of preventative versus positive checks to population growth.

Both Malthus and Clarke had concluded their observations on the state of modern Swedish society that government involvement in the promotion of science and medicine left much to be desired. For Malthus, government collaboration with the *Collegium Medicum* to assist population health and establish institutions to care for the ailing and disabled was unfair to the rest of the population, and he believed that government resources could be better spent on encouraging the healthy in more productive pursuits. Clarke believed that the government's early framework for practical training and systematic education helped Swedish science to flourish in the eighteenth century. By the century's end, however, the loss of patronage, shift in science's perceived economic function for the state and changing ideologies of education had stunted scientific and technological progress there.[81]

But Scandinavia was a large land mass, and both Malthus and Clarke were anxious to gather information about a very different realm of life in the north – one entirely removed from the scope of their critiques of state support in science and medicine. In the mountains, forests and marsh lands of the extreme north lived another population that the British knew little about. The Laplanders were a people enthusiastically sought out by Malthus and Clarke, and the further north they travelled in pursuit of this nomadic population, the more they struggled to make sense of the distinct and unfamiliar way of life in the 'frozen north'. The Lapps, the aboriginal population of northern Scandinavia who lived on what was judged the 'border of European civilisation', were subjects of an anthropological analysis used to explore the contrasts between the civilised and barbaric.

The 'borders of civilisation'

In the eighteenth century, an image of the northern lands of Scandinavia as a mysterious place emerged in an array of literature. Accounts were written of witchcraft and wizardry amongst the Lapps, who purportedly possessed magical powers and practised sorcery by beating magical drums. Allegations were made that the Lapps were conjurers, who peddled evil spirits, were merchants who sold wind and storms, and cast spells on their southern neighbours. Many references to Lapland wizardry were inspired by Johannes Scheffer's *Lapponia* (1673), enlarged and translated as *History of Lapland* (1674, and subsequent editions), which became the authoritative text on that region throughout the century.[82] As one late eighteenth-century writer commented, drawing from Scheffer's earlier account, 'There is scarce a country under the sun, where the name Lapland has reached, but what has heard of their magic, which is not quite abolished among them.'[83] In English periodicals such as the *Spectator* and the *Gentleman's Magazine*, streams of poetry, songs, legends and commentary about northern superstition were published.

In Daniel Defoe's *A System of Magick* (1726), the Lapps were associated with sorcerers and soothsayers who were in the service of the black arts. The

> Wind Merchants in *Norway*, who sell fair and foul Weather, Storms and Calms, as the *Devil* and you can agree upon a Price, and as your occasions require: Also in *Lapland, Muscovy, Siberia* and other Northern Parts of the World, he is said to act by differing Methods, and govern his Dominions by a more open and arbitrary Method, not prescribed and limited to Art and Craft as he does here.[84]

Other writers were more sceptical of the myths and legends that were spun out in discussions of Lapland. In 1726, the poet James Thomson published *Winter*, which moved away from their magical powers and portrayed the Lapps as noble savages who lived in a picturesque environment. The wandering Lapps lived a primitive life but none the less adapted to their surroundings and succeeded in living a life of comfort and complacency.

> Not such the Sons of Lapland: wisely They
> Despise th'insensate barbarous Trade of War; . . .
> Thrice happy Race! By Poverty secur'd
> From legal Plunder and rapacious Power:
> In whom fell Interest never yet has sown
> The Seeds of Vice.[85]

Romantic literature played up the motif of the primitive northerner adapting to the harsh climate. Their nomadic freedoms complemented the sublime representations of the landscape: they moved with the seasons over dynamic, picturesque terrain. In John Campbell's *Polite Correspondence: or, Rational Amusement* (1740), two fictional travellers exchange letters, one writing during a five-month tour through Lapland. He relayed his observations that the Lapps were not nearly 'so wild or so bad as they are generally represented', that despite the conditions they live adequately, and that their contentment overpowered a desire to seek more comfortable lives. 'I agree entirely with *Philintus* in acknowledging the Injustice done the *Laplanders*, while we rail at them as a base and barbarous People', wrote Florimound, another of Campbell's fictional correspondents.

> For how, my Friend, can a nation be called *Base*, who are so remarkably fond of *Liberty*, or why should we stigmatize as barbarous, Men, who live agreeable to the Soil and Climate where God and Nature have plac'd them. A Sledge is certainly a more convenient Vehicle in *Lapland*, than a Coach and Six, and it is therefore Folly in us to blame them for using it. We ought rather to admire the Wisdom and Goodness of Almighty God, who hath so exactly suited the Genius and Temper of all nations to the Circumstances attending those Regions, wherein according to the Disposition of his Providence, they are to inhabit. Neither is this less admirable, though it should appear that the Genius of a People is in some measure govern'd by their Soil and Climate, since this affection must be produc'd by certain laws impressed by God or Nature.[86]

It was an argument from design; God's infinite benevolence and wisdom had made the Lapps' customs and means of subsistence perfectly adapted to the northern environment. Campbell's fictional commentators suggested a providential resonance between the northern landscape and the characteristics of the inhabitants.

More philosophic than fictional writers contemplated what would induce anyone to live in such a severe climate. Some reasoned that the Lapps were inured to the conditions, and through experience and familiarity learned to conform to the circumstances. Tobias Smollett found nothing irrational or unnatural about those born in a place developing local attachments towards the land. A feeling of attachment seems to be a kind of fanaticism founded on the prejudices of education, which induce a Laplander to place the terrestrial paradise among the snows of Norway, and a Swiss to prefer the barren mountains of Solleure to the fruitful plains of Lombardy. I am attached to my country, because it is the land of liberty, cleanliness, and convenience: but I love it still more tenderly, as the scene of all my interesting connexions . . .[87]

But these were musings based on scant knowledge and experience of life in the north.

Few scientific texts existed which added empirical data to the stock of knowledge about the resources and geography of Lapland. Carl Linnaeus's *Flora Lapponica* was published in 1737. This was a compilation of the northern flora and fauna that Linnaeus encountered during a summer excursion to Lapland in 1732. The trip was profitable for Linnaeus; it established him as a competent natural historian and collector, and at subsequent ceremonies and during Continental travels he would don the Lapp costume, complete with a biretta (woman's headgear) and a drum. The next year, Pierre Louis Moreau de Maupertuis published *The Figure of the Earth* (1738), the report of the northern expedition to measure the arc of the meridian. This included maps, illustrations of living conditions (mainly of the Swedish and French explorers) and spectacular accounts of the northern lights.[88]

These accounts were useful, but even so, it was easy to sense a void in the knowledge of Lapp community, customs and moral well-being. Over the centuries, missionaries from Norway, Sweden and Russia had travelled to gather information and introduce Christianity to the Lapps. Spurred by rumours of paganism and mysticism, the strongest attempts to evangelise the north were undertaken in the early eighteenth century. Missionaries were then required to have an intimate knowledge of Lapp customs and language; the New Testament appeared in the native tongue in 1755, and the Lapp language was codified in the *Lexicon Lapponicum* (1780).[89] One English visitor to Lapland who worried about the cultivation of religion there was Matthew Consett, who travelled in 1786 (accompanied by Sir Henry St George Liddell and a 'Mr Bowes'). On his way from Stockholm to the north he wrote that upon 'leaving behind us these traces of civilized life we entered into woods that did not terminate for many miles'. The gallant trio soon entered *terra incognita*, surrounded by reindeer and wild fowl. After a tour through 3,784 miles of northern territory in the space of fifty-odd days, they decided to return to civilisation, with Consett hoping that the Laplander's Christian neighbours would proselytise these 'yet unenlightened People'.[90]

The northern frontier

Heading north through Sweden on their way to Lapland thirteen years after Consett, Clarke and Cripps received cautions about what they would encounter in the 'Frigid Zone'. They were told that the 'largest bugs in the world would attack us in Lapland' and Clarke recorded that the Swedes still thought the Lapps were witches or magicians, possessing the power to commit injuries upon people at a distance.[91] While intrigued, it is easy to imagine how anxiety, or even trepidation, could exceed mere curiosity about what they might find above the Arctic Circle, their ultimate destination.

The eager travellers made rapid progress north along the coast, and in early July they were near Umeå, in Västerbottens – part of the interior of which was politically designated Lapland territory. They were already engaged in a unique experience. The summer solstice had just passed, and the Englishmen found it difficult to deal with the midnight sun. Having travelled all day, they stopped just after midnight and wrote in their journals, without the aid of candlelight. They sat and soaked in the reflections of the relentless sun, fatigued, but the brightness prevented them from sleeping. Clarke determined that darkness was a benevolent gift of Providence, the value of which was the pleasure of repose. The mode of transportation was uncomfortable, but just as other travellers had, Clarke admired the efforts and dedication of the peasants in northern Sweden to maintain the roads.

He was also struck by the physique of the northerners. Restless, he imagined what notice would be attracted by Londoners if these Swedish peasants – with their strong, 'gigantic statures', tanned skin and pearly teeth – were seen with their long, flowing locks of hair, barefoot, commanding their carts along St James's Street or Hyde Park.[92] The displacement felt by these English gents, resting in the bright, birch-filled countryside of northern Sweden, was inverted in the fantasy of seeing the local peasants gallop through London. Clarke contemplated his and his companion's curious experiences in a most self-conscious way. In order to demonstrate the striking contrast between the usual and the unusual, he invited his readers to imagine encountering the unexpected in their familiar world. In an inversion of anthropological technique, he jostled his readers into thinking about the traveller's experience by juxtaposing their routine lives against the intrusion of the foreign. The further north they travelled, the more unfamiliar things became. The Swedish language began to alter and their interpreter began to have difficulties communicating. They collected trinkets as presents for the Lapps and began to search for specimens illustrated in Linnaeus's *Flora Lapponica*. Then, in Luleå, 'to our great satisfaction, we saw, for the first time, some of the *Laplanders* in their native dresses'. Clarke was overjoyed. 'A Lapland woman, attracted by curiosity, came, with her husband and child, into the room where we were getting some refreshment: and such was our delight upon seeing her, that, ugly as she was, we even ventured to kiss her; a liberty she did not at all seem to approve.'[93] This was Clarke's first opportunity to examine closely the attire and accoutrements of a Lapp family (see Plate 5).

Their attire was much the same design as it had been for centuries. Garments were typically a bluish colour, made of sheep-skin and decorated with colourful red and yellow cloth ribbons sewn to the shoulders. Some coats were embroidered with tin threads, with thick woollen sashes or black ox leather belts. Their pointed shoes were made of reindeer chamois. Rather than stockings, the Lapps lines their shoes or boots with soft, combed sedge grass, capable of drying quickly.[94]

They had just entered Lapp territory, although it was difficult to be exact about boundaries. Clarke wrote that 'Charles the XIIth, whose policy directed him to preserve the Laplanders from mixing with the Swedes, sent engineers, in 1690, to mark, with all possible precision, the southern frontier of Lapland. Still, however, they are indeterminate.'[95] From the Middle Ages, the hunting grounds of the Lapps had been aggressively reduced by Norwegian, Swedish, Finnish and Russian colonisation. In the eighteenth century, the Lapps were allotted land rights, the privilege of which being susceptible to taxation, levied by the neighbouring governments (sometimes doubly for the same bit of land given competing territorial interests by the northern countries). In 1751 the mountainous borders between Norway and Sweden were established, with provinces extending down the middle as far south as Røros, extending to the coastal regions only in northernmost Norway. About the same time, the 'Lappmark frontier' in Sweden was drawn approximately 60 miles inland from the coast, and this was to protect

Swedish agricultural interests from excessive Lapp migration. In 1809, when Russia conquered Finland and the adjacent parts of northern Sweden, and in 1826, when the northern Norwegian–Russian borders were redrawn, Lapp territory was again affected. Further geo- and ethnographic divisions occurred within the Lapp population, representing 'Norwegian Lapps', 'Finnish Lapps', and so on, with slightly different customs and language. To this day, the boundaries and cultural autonomy of the Lapp (or Sami) population are unsettled issues.

Clarke's encounter with the Lapp family was in early July. Close to a month later, and much further south in Norway, Malthus and Otter approached 'Roraas' (Røros), looking for Lapp territory. 'Over all these mountains that separate Norway from Sweden', wrote Malthus in his travel diary, 'we understand, are a number of wandering Laplanders who are sometimes on the Swedish & sometimes on the Norway side of the mountains'.[96] Malthus and Otter travelled from farmhouse to farmhouse, seeking information on the best way of finding the Laplanders, whose nomadic lifestyle, especially during the summer, made them elusive targets. Finally, however, they heard of a sighting. 'Much pleased with this information we pushed forwards, & when we were getting out of the woods looked with great eagerness round in search of Laps [*sic*] – the man who first view'd one, was to have 6 pence.'

On horseback, they had rushed beyond birches, over barren hills, through valleys, past lakes and up to the mountains. Their eyes were fixed on the horizon, scanning for the nomadic people. Then, they heard dogs barking and soon after saw huts. Finally, upon approaching the temporary settlement, Malthus witnessed a 'she Lap' peep out from a hut and he triumphantly claimed his prize. During this contest, or pseudo-sporting event, Malthus talked of gathering 'intelligence' in their 'pursuit' that might help to 'find', or 'catch' the Lapps.[97] One can imagine the burning curiosity these ordained English gentlemen experienced when they dismounted and evaluated the living conditions before them. The Lapp huts were constructed of intertwined branches, covered with cloth, just tall enough for the '5 foot fins'. Inside, the children were found 'sleeping quite naked' on reindeer hides.

Malthus's encounter afforded him the much desired opportunity to record details of the appearance, customs and language of the Lapps. But his time was brief – only two days – and his notes occupy just seven printed pages. His observations mainly concerned daily habits, and his inquiry seems to have been somewhat regulated by the discrimination of their translator: 'There were many more questions that we wished to ask', lamented Malthus, 'but our servant naturally thought so many questions trifling & foolish & shewed a little unwillingness to repeat them'.[98] Clarke, however, was able to venture further into Lapland territory, and his was a more extensive account.

The first lengthy discussion regarding Lapp culture that Clarke brought to the attention of his readers concerned their language. The study of language played an important part in comparative cultural analyses.[99] For some, possessing a rational faculty of language was thought to make the emergence of 'society' possible, and it became a characteristic in classification schemes which separated some 'anthropoid' species from others. Besides questions of the separation of species, however, the comparative study of language was also used to trace the historic origins and connections between different groups of people.

In *Ideas for the Philosophy of the History of Mankind* (4 vols, 1784–91), the German philosopher Johann Gottfried von Herder argued that a comparative study of different cultures was crucial to understanding laws of human nature, and that through the study of a nation's language one could obtain intimate knowledge of the values of each culture. In other essays, Herder's belief that the function of language revealed intimate aspects of human life led to the study of German music, Norse poetry and mythology, and Slavic folklore. Complementary literary studies, such as those by the German philologists Jacob and Wilhelm Grimm on myths, fairy-tales and folk customs, were believed to bring about a new understanding of a people's national identity. Attaching a study of language to analyses of national identity were also often studies in human history and diversity. In Sweden, the emergence of 'runology', the study of runic script and symbols, also fed narratives of a nation's historical identity.[100]

Swedish antiquaries had also studied structural and phonetic aspects of Lapp language to shed some light on their cultural heritage. Clarke referred to one theory that Lapp ancestry was linked to the Hebrew tribes who formed the ancient Kingdom of Israel. This refers to the story of the conquest of the promised land by the Assyrians in 721 BC, when the ten original tribes were assimilated by other peoples and 'disappeared' from history. Swedish antiquarians suggested that a portion of these people migrated to the north, settled in Lapland and 'bestowed their own appellations upon the mountains, lakes, and rivers'. It was further suggested that even in the eighteenth century the 'Lapland language approaches near enough to the *Hebrew* for the two people to understand each other's speech'.[101] Clarke was sceptical, but his summary of the debate about the Lapp language suggests an association between the philosophy of the history of mankind and emerging studies in comparative philology, such as the British orientalist Sir William Jones's. His work in the 1770s and 1780s revealed a strong structural affinity between Sanskrit and Greek and Latin, leading to the conviction that these languages must 'have sprung from some common source'. With this linguistic relationship came the possibility of a shared cultural ancestry, though such claims were resisted by German and British scholars who rejected ideas that their languages were the result of 'impure mixtures'.[102] Clarke's concern to address debates about Lapp language and history should be seen in the context of these wider debates in what we now refer to as 'comparative ethnolinguistics'. However, Clarke believed that lack of thorough investigation allowed only preliminary speculation in this area. All these debates proved, he suggested, was 'how very little is yet known respecting the origin of this singular people'.

If one added to language analysis an investigation into other physical characteristics and customs, however, then theories about Lapp cultural origins could become more convincing. For Clarke, the 'absolute certainty of an *Asiatic* origin of the *Laplander* is conspicuous in all that belongs to his person, in his complexion, pliant postures, diminutive stature, air and manner...'.[103] His physical description of the Lapps in order to evaluate their cultural circumstances was in part bound to earlier judgements of the Lapps 'primitive' status.

But the 'primitive' itself was a classificatory term which involved historical perceptions of the civilising process. Thus, exploring the extreme boundaries of 'civilisation' was partly motivated by the preoccupation to apply categories to undefined 'primitive' specimens. The eighteenth century found many 'human like' animals in appearance, but their language and actions were of a bestial order.

The pervasive eighteenth-century belief in the 'chain of being' – the concept that all natural specimens were 'chained' together in a hierarchy according to increasing physical complexity and mental sophistication – was one framework for a classification scheme. Frequently, however, it was difficult to determine where in the hierarchy a specimen should be placed. After the 'discovery' of African apes and their dissection by the French comparative anatomist Georges Cuvier, naturalists were shown striking similarities between human and ape bodies and brains. Such close ties were exemplified in Linnaeus's famous classification scheme in the 1758 edition of *Sytema Naturae*. Here, for the first time, humans were grouped with apes in the same order (primates) and even the same genus (homo) as orang-outangs. Linnaeus's designation of humans in the species *sapiens*, meaning 'wise', was a gesture towards making man an essentially superior animal.[104] Humans were wise (sapient) and rational, as defined by European standards. This category was therefore problematic to naturalists looking to classify 'human-like' savages.

The case of Peter, the 'wild boy' of Hanover, who was abandoned by his parents in the 1720s, spawned popular discussion about the civilising process and the degrees of difference between rational humans and noble savages. After he was found in the forests and returned to society, he was taught to dress, gesture and compose himself as a civilised man, and was displayed at the court of George I, though he never acquired language skills. Subsequent similar human studies also contemplated the power of human improvement: Victor, the 'wild boy' of Aveyron about 1800; Kaspar Hauser in the 1820s; and 'Jemmy', the Fuegian savage, who was educated and returned to South America in 1832 on board the *Beagle* with Darwin, who was horrified upon his later return to Tierra del Fuego to find Jemmy's civilising mission had failed, and he himself had degenerated to 'so complete & grievous a change'.[105]

Eighteenth-century discussion of the Lapps resonated with these classificatory dilemmas. The Russians referred to their rights of trade and taxation on the 'wild Lapps' (Dikaia Lop), an expression echoed by Clarke. But such references showed carelessness as much as confusion. Clarke combined observations of the language with physical traits in attempt to gather enough information to reach a reasonable conjecture about their history. Questions concerning the Lapps' 'origin, hitherto not developed, would afford one of the most curious subjects of inquiry hitherto offered for consideration, as affecting the history of the human race,' he wrote.[106]

The Lapps, he observed, had distinctive physical characteristics. They were short, had straight, dark hair, lean and swarthy faces, prominent cheek bones, short necks and narrow eyes sunk into their skulls. 'The whole race of *Laplanders* are pigmies,' he asserted, which again evoked Linnaean attributes of feral man, or (earlier)

Edward Tyson's *pygmy*, a young chimpanzee, which for other writers came to represent different forms of 'humanoid apes'.[107] The nomenclature was as indeterminate as the criteria for classification. Clarke endeavoured nevertheless to situate the Lapps into some kind of human classification scheme. He was committed to the idea that they were *special*, while part of his broader genus. 'If it were granted, that *man*, like other animals, admits of being distinguished into many separate species, we should not hesitate in considering the genuine *Lapp* as one of these.'[108] While spirited, if also controversial, attempts were made in the eighteenth century to draw conclusions about the classification of humans and animals, it was only in the nineteenth that anthropological, medical and evolutionary assessments were used to refine classification systems. The early nineteenth-century work of the Bristol physician James Cowles Prichard on the geographical distribution of racial groups is, however, interesting to note. His researches in physical anthropology opened up new areas for comparison and controversy, but his thoughts about distinct human races could almost have been written had he travelled with Malthus and Clarke:

> Providence has distributed the animated world into a number of distinct species, and has ordained that each shall multiply according to its kind, and propagate the stock to perpetuity, none of them ever transgressing their own limits, or approximating in any great degree to others, or ever in any case passing into each other. Such a confusion is contrary to the established order of Nature.[109]

But given that Prichard wrote this in his *Researches into the Physical History of Man* (1813), it is conceivable that his arguments for distinct species might have informed Clarke's discussion of the Lapps in his published *Travels*, since he composed that volume in 1818.

Forging classification systems raised questions about physical and social similarities and differences, and about essential human traits as opposed to regional, environmental varieties. Even with increasing concerns to map geographical distinctiveness, this was not prioritised over a chronological model of social development. Even with his principal concern to record the details of their present condition, Clarke reserved a space for a swift estimation of their social structure. In Malthus's terms, their pastoral, nomadic way of life more resembled barbarian shepherds of past epochs than anything remotely modern. Clarke noted that the Lapps in 'Tornea' (Tornio, at the apex of the Gulf of Bothnia, on a trading path) were 'not well versed in commercial speculations; if they were, they might soon become rich'.[110] As it was, their wealth was determined by how many reindeer they possessed – the benchmark of their barter system. This was to be expected of a community stuck in a 'young', 'pastoral state', as defined by the 'conjectural history' of the stages of civilisation. 'The manufactures of a people in such an incipient state of society are, of course, little worth notice,' thought Clarke.[111]

Observations on their language, coupled with their physical appearance and social evolution, presented a somewhat qualified, but none the less speculative, assessment of the Lapps' position amongst the human races. Their lives in the distant northern lands, 'cut off from all communion with society; whose dwarfish stature, and smoke-dried aspect, scarcely admits of their being recognised as intellectual beings "created in the image of God"'. But their primitive lives were waiting to be explored further. What Linnaeus had once described as a tranquil life of unbroken solitude was ripe ground for scientific study, 'if a man presumes to penetrate'.

> What then are the objects, it may be asked, which would induce any literary traveller to venture upon a journey into Lapland! Many! That of beholding the face of Nature undisguised; of traversing a strange and almost untrodden territory; of pursuing inquiries which relate to the connexion and the origin of nations; of viewing man as he existed in a primaeval state; of gratifying a taste for *Natural History*, by the sight of rare animals, plants, and minerals.[112]

It was exactly the strangeness, its singularity and the lack of knowledge about the land and people that made travelling there consequential for the literary traveller.

An alternative Enlightenment?

Travellers' journeys through Scandinavia generated much material with which British commentators could begin to evaluate the condition and boundaries of 'European' civilisation. The centrality of Sweden in their observations reflected both the concern to chart geopolitical securities in a Europe struggling to maintain a proper 'balance of power', as well as their interests in re-evaluating the enlightened achievements of a country with a dynamic history of political and social reform. Two moments in Sweden's history were identified as turning points in the road to their Enlightenment: after the death of Charles XII in 1718, Sweden's military prowess began to weaken; after the assassination of Gustav III in 1792, it seemed that their intellectual and cultural fitness was also doomed to decline. On the eighteenth-century measure of modernity, Sweden was sliding down the scale.

Nevertheless, the Baltic region did provide historical lessons regarding how enlightened theories might work towards the making of a modern society. Politically speaking, Gustav III's 'bloodless revolution' of the monarchy did provide a stark contrast to the horrors of the revolution of the 'third estate' in France in 1789. Chaining and publicly displaying the rogue citizen who shot Gustav III seemed justified; few in Britain would declare it just that France's Louis XVI was publicly decapitated in 1793. At the century's end, some worried British commentators were keen to blame the radical free-thinking French *philosophes* for the misguided revolt against the God-given right to rule. But didn't Sweden give birth to some of the most celebrated and pioneering natural philosophers of the century? And didn't they lay the foundation for an enlightened and modern system of scientific rationality, admired throughout Europe?

Not everyone was willing to condemn natural philosophy as an inherently subversive enterprise. Indeed, Sweden seemed to prove that through proper state support, natural philosophers could provide useful information about, and bring order to, the natural resources upon which national economies relied. Not only did Sweden's (as well as Norway's) state apparatus rely on the exploitation of its natural resources that were mined and defined by her civil servants, but, in terms of foreign trade, so too did others – such as Britain. Especially after 1793, when Britain's relations with France turned particularly hostile, the Baltic States were of special interest. This interest was political, military and economic. As Clarke pointed out, it was through the patronage of the 'revolutionary' Gustav III that intellectual life – let alone artistic opulence – was given its last breath. It was through proper, state-supervised training that the 'true foundations of science' were laid, and through which science could best be used in the service of the state. By the end of the century, the concern became whether or not those foundations would prove strong enough to support the proper balance of power in Europe.

Not everyone agreed on the proper role the state should play in patronising the sciences and 'protecting' its population. This was seen, for example, in Malthus's comparative analysis of agricultural and medical innovations in Sweden and Norway. But for both Clarke and Malthus, and indeed for most travellers who explored Scandinavia, the observations accumulated through travel provided not only the basis for assessment of how well 'modernity' was managed, but the raw data and empirical insights for what might work in British society.

The 'limits' of Enlightenment in Scandinavia were not only defined relative to a political and intellectual landscape, but according to various ways of classifying populations at the geographical frontiers. In juxtaposition to the high grounds and pitfalls of modern political and intellectual life in the European north laid the land of the 'as yet unenlightened' Laplanders. Here were yet more resources for population analysis – areas where new analytical tools could be deployed to discern contrasts between 'European', 'extra-European' or 'non-European' peoples. In the 'frozen north', information on climate, natural history, diet, migration, clothes, language and physiological appearance was gathered in an attempt to explore the possibilities of some shared cultural ancestry amongst populations at different areas in the European frontier. Of course, any attempt to classify largely unfamiliar populations was riddled with prejudicial assumptions, and always left more questions about the similarities and differences between populations than it answered. In the eighteenth-century frame of mind, such lack of understanding was a motive force for going further into the frontiers, accumulating more data and working towards more comprehensive comparative analyses.

Such was on Clarke's mind when he and his travel companion left Scandinavia and headed into Russian territories. The comparisons between the different northern countries and Lapp territories, which was observed with reference to different interests in agriculture, manufactures, pastoral farming, manners, customs, and so on, proved a lucrative start to a European tour. However quickly they had traversed Scandinavia, the travellers wasted no time pushing ahead to new frontiers.

. . . Into Russia . . .

'We know not how to paint the extreme contrast which appears in the short distance of an *English* mile – from the *Swedish* to the *Russian* guard. The country is still *Finland*, but it is *Russian Finland*; and to heighten the differences between an union with *Sweden*, and a subjugation by *Russia*, the *Russian Finns* are not those who make their appearance at the guard, but soldiers from the interior of the empire; the reason of which will soon appear. In a company of the *Tavasthuus* militia, stationed at a small distance from the *Swedish Douane*, on the east side of the western branch of the river, which separates the two countries, we had the last view of the benevolent and mild inhabitants of *Sweden*. They were a sturdy and athletic troop: and as it gave us a melancholy satisfaction to prolong the few moments of our farewell, by conversation with them, the officer on duty politely accompanied us as far as the *Russian* guard.

In passing the little island which lay between the *Swedish* and the *Russian* bridge, we expressed a curiosity to know what formed the precise boundary of the two countries. The *Swedish* officer shewed us a stone of about two tons weight, which is the only object that is supposed to break the neutrality of this interval between the respective posts. Higher to the north is the *Tammijara*, a small lake in the western branch of the *Kymene* River; which river, with the more remote waters of the *Pyhå* and *Wuokå* lakes, forms the line of demarcation . . .

A few miles away, nay, even a few yards, conduct you from a land of hospitality and virtue, to a den of thieves.

We suffered from this want of caution, in the loss of the first moveables on which the *Russians* could lay their hands. We had, indeed, been forewarned of their pilfering disposition, but did not imagine that we should so soon experience the truth of the information which we had received respecting this part of the *Russian* character.'

Edward Daniel Clarke, *Travels in Various Countries of Europe, Asia, and Africa* (London, 1824), Vol. 11, pp. 372–4

3
Eastern Frontier: Russia and its Frontier – Cultivating Empire and Imitating Enlightenment

The boundary between Swedish and Russian territory – represented by a 'stone' – seemed to Clarke an insignificant physical point to pass when compared to the contrast in character between the 'sturdy and athletic' Swedes and the 'den of thieves' which, Clarke told his readers, he immediately encountered in Russia. As we will further see, Clarke's was a harsh account of the manners and conduct of the government and nobility in metropolitan Russia. His observations of the customs and condition of life in the eastern and southern provinces rounded off a portrait of an empire with most of its expanding population no more civilised than the Lapps.

Because the acquisition of knowledge about Russian society was sluggish throughout most of the eighteenth century, any views, however controversial, drew attention to themselves. As a growing European government, an ally against Napoleon, an increasingly important trading partner with the West and a rapidly emerging power with an expanding empire, Russia was overdue for a place in comparative analysis, exploration and first-hand travellers' reports. It was not an easy empire to get a clear focus on. Its boundaries had shifted many times throughout the eighteenth century and its constituents changed with every military conquest.

Bordering a lack of knowledge

No unproblematic series of dates can be cited for when Russian territories and borders took shape during the eighteenth century. The acquisition of new territorial tracts or the military enforcement of boundaries demarcating separate sovereign states were variously established, changed, debated and reassigned through numerous diplomatic contests. The Russian frontier fluctuated. But while exact boundaries might have been difficult for eighteenth-century visitors to Russia to perceive, what was undeniable and most discernible was the empire's breadth of expansion in all directions.

Russia's great land mass, extending from Europe into Asia, led to a distinction occasionally made between the 'two Russias'. Where the two continents divided changed throughout the eighteenth century. Sometimes the Don, sometimes the Volga, separated 'European Russia' from 'Asian Russia'. Other times the Ural Mountains formed the physical boundary, as they generally do today.[1]

An expanding empire

Significant territorial acquisitions were made during the reign of Peter the Great (Peter I). In 1689, the Treaty of Nerchinsk was signed between Russia and China, establishing a long-standing north-eastern border. In Siberia, a forced separation between submissive and hostile natives regulated the frontier.[2] The steppes of the south-eastern frontier were just as loosely defined. In 1696 Peter I captured Azov from the Crimean Tartar vassals of Turkey, which allowed him to claim territory which extended to the Black Sea. In 1723 Persia ceded the western and southern shores of the Caspian to Russia. The south-western frontier, from Azov to Kiev, was secured by maintaining a chain of military fortifications. From 1667 Kiev and left-bank Ukraine (the land east of the Dnieper) were relatively autonomous and under protection of Moscow, but in 1686 Kiev was brought further under Muscovite hegemony, which started a process of Ukrainian annexation, leading to the assemblage of what was called 'Little Russia'.[3] Throughout the eighteenth century, the Ukraine (the name itself meaning 'edge' or 'periphery') gradually lost its independence, and in 1793 Kiev was fully incorporated into Russia.[4]

When Peter I died in 1725, the Russian empire stretched from Archangel on the White Sea to Mazanderan on the Caspian, and from the Baltic Sea to the Pacific Ocean. During the reign of Catherine the Great (Catherine II), the empire expanded further. As a result of the Russo-Turkish war (1768–1774), Russia won the right to free navigation in the Black Sea and was able to colonise its northern coasts, increasingly annexed from Turkey between 1774 and 1792. In 1783 Catherine annexed the Crimea, adding a significant appendage to 'New Russia' and providing a further blow to the body of the Ottoman Empire.[5] Also during this time, an agreement struck between Catherine, Frederick II of Prussia and Austria resulted in the partitioning of Poland three times, in 1773, 1793 and 1795, giving Russia control of Belorussian lands. At the end of Catherine's reign (1796), the Russian empire had expanded westward (over most of the western Ukraine, Poland and further along the Baltic coast) and southward (to the Caucasus Mountains and the Caspian Sea). The empire covered 200,000 square miles of European and Asian land, had the largest population of all the European states (estimated at almost 26 million in 1783), and boasted the longest river on the continent (the Volga).

The rate of imperial expansion did not go unnoticed by European contemporaries. 'The eighteenth century was opened with a very interesting scene for all the northern parts of Europe,' noted the political historian John Williams.[6] The territorial tensions in the north, particularly with regard to the long-standing contests between Sweden and Russia, created an interesting scene because of the shift in the balance of power in the eighteenth century. The century began with the Great Northern War (1700–21), which lasted almost throughout Peter's reign. The war started with an anti-Swedish coalition of Denmark, Russia and Saxony-Poland attacking Charles XII of Sweden. When hostilities ceased, Sweden was able to maintain possession of Finland, but lost most of their extended eastern Baltic empire to Russia, whose success in this war established it as an important player in European politics.[7]

Eighteenth-century Russia eclipsed Sweden's seventeenth-century position as an imperial power. Along with growing military might and political prowess around the Baltic, Russia was developing continental interests and increasing trade with Britain and other European states. Peter promoted interest in the mining and metallurgical industry, and pig-iron production from the Urals grew exponentially. Eventually Russia surpassed Sweden in exports of raw materials such as timber and iron to Britain.[8] By the later eighteenth century, Britain, as well as other European states, was realising what little knowledge had been systematically collected about this imposing empire. The Russian frontiers were cloaked in mystery, slowly revealing their riches to the rest of the Continent.

An expanse of inquiry

William Tooke, chaplain to the Russia Company (the 'British Factory' as it was called in St Petersburg) from 1771 to 1792, wrote about Russian aggrandisement. 'Russia, an empire but little known or regarded in the last century, at the opening of the present made her appearance all at once among the states of Europe; and, after a short trial of her power, became the umpire and arbitress of the North.'⁹ His *View of the Russian Empire* provided extensive observations on the natural resources of the territory and the diversity of its people. He provided a summary of the frontiers that were fixed by treaty, citing agreements with China, Persia, Poland, Sweden and the Turks, but admitted that the borders were sometimes ambiguous and 'not yet settled'. Russia's relentless attempts to expand its borders caused fear in the north, anxiety in the south and curiosity in Western Europe. Their territorial contests confused others, who wondered why the natural borders did not work to define national boundaries, such as the Baltic in the north and the Don in the east. 'What madness then urges Russia, Sweden and Denmark to worry each other?' asked Andrew Swinton, a Scottish traveller who toured the northern countries just as the French Revolution erupted. 'Empires, like Individuals, have their family pride,' he answered, and so rulers did not forget the means by which they were able to inherit their kingdom and ensured that their armour did not rust.¹⁰

While the shifting frontiers and accumulating strength of the Russian empire captured the attention of increasing numbers of travellers and political commentators, their impressions were by no means uniform. Due to the lack of accumulated knowledge and the propagation of careless speculation early in the century, portraits of Russia remained as ambiguous as its borders. But what made the activities and observations of late eighteenth-century travellers significant was the ambition with which they gathered factual knowledge: they mapped, recorded and collected information in a range of specialised fields of inquiry.

But, compared to the resources tapped by the French and Germans, the British were lagging in the pursuit of knowledge about the Russian empire. The Russian frontiers were labelled 'unknown territory' on maps sold in London until the latter half of the century, and even the improved were riddled with errors, as many enterprising travellers subsequently discovered. Geographical knowledge about Russia came in dribs and drabs, through occasional translations from foreign authors, such as the Frenchman Cornelis de Bruin's *Travels into Muscovy* (1720), the journal of the Dutch minister F.C. Weber's *The Present State of Russia* (2 vols, 1722–3), or the Swede Philipp Johann von Strahlenberg's *An Historico-Geographical Description of the North and Eastern Parts of Europe and Asia* (1738).[11] In mid-century, information about Russian geographical expansion came particularly from German travellers who were recruited by the Academy of Sciences in Petersburg. These included Johann Gottlieb Georgi's *Beschreibung aller Nationen des Russischen Reiches* (1776), Gerhard Friedrich Müller's *Voyages from Asia to America for completing the discoveries of the north-west coast of America* (1761), the geographer from Tübingen Johann Georg Gmelin's *Flora Sibirica, sive historia plantarum Sibiriae* (1747–69) and (among his numerous publications) the German naturalist Peter Simon Pallas's *Travels through the Southern Provinces of the Russian Empire, performed in the years 1793 and 1794* (1802–3).[12] Despite their descriptions, it appears that cartographic knowledge remained neglected.[13] As late as the 1800s, when Clarke was preparing his *Travels* for publication, he found that a recent map published in Berlin (1802) was useless, but attributed this to the Russian policy of suppressing information about their territories:

> the Don Cossacks, Kuban Tartary, and the Crimea, appear [in the map] as a forlorn blank. Many years may expire before *Russia*, like *Sweden*, will possess a HERMELIN, to illustrate the remote geography of the provinces of her empire; especially as it is a maxim in her policy, to maintain the ignorance which prevails in *Europe*, concerning those parts of her dominions.... The only tolerable charts are possessed by the Russian Government, but sedulously secreted from the eyes of Europe. It has however fallen to the author's lot, to interfere, in some degree, with this part of its political system, by depositing within a British Admiralty certain documents, which were a subsequent acquisition, made during his residence in *Odessa*. These he conveyed from that country, at the hazard of his life.[14]

British observations became more acute during the last two decades of the century, partly due to feats of espionage akin to Clarke's furtive achievements. Tooke began his two-decade, comprehensive survey of the Russian empire by translating Georgi's *Beschreibung* as *Russia: or a Compleat Historical Account of all the Nations which Compose that Empire* (4 vols, 1780–3), upon the completion of which he was elected Fellow of the Royal Society in London. Between 1785 and 1787, a series of letters sent by Tooke during his voyage down the Don to the Sea of Azov were published in the *Gentle-man's Magazine*, some of which provided the first descriptions of the Crimea to the British public.[15] His *View of the Russian Empire* (mentioned above) and some additional translations were followed by his *History of Russia from the Foundation of the Monarchy by Rurik to the Accession of Catherine the Great* (1800). Tooke's voluminous contributions to the annals of Russian imperial history made him a distinguished scholar, although he granted that the expanding empire left correlating expanses of knowledge to be harvested.

Exploring the different cultures of all the people who were absorbed into the Russian empire was a demanding task. More was left to be learned about the effects of the 'rapid and violent' introduction of foreigners to imperial rule. What were the effects of extensive intercourse with new populations from the frozen north to the Adriatic, from the shores of the Neva to the Pacific? Most nations of Europe, Tooke observed, have their own uniform customs, common physical features and language. Not so in Russia. There 'dwell not only some, but a whole multitude of distinct nations; each of them having its own language, though in some cases debased and corrupted, yet generally sufficient for generic classification; each retaining its religion and manners, though political regulations and a more extensive commerce produce in some a greater uniformity.'[16] The late eighteenth-century Russian population was indeed diverse: it contained some 40 per cent Ukrainians or Belorussians, 26 per cent Poles, 20 per cent Lithuanians, 10 per cent Jews, and 4 per cent Russians; 38 per cent were Catholics, 40 per cent Uniate, 10 per cent of the Jewish faith and 6.5 per cent Orthodox.[17] To the eighteenth-century historian John Williams, this diversity made Russia distinct to Europe: 'In all the other states of Europe, the people who live under the same government are very little different one from the other, but this is not the case with all the people who compose the great Russian empire, where the inhabitants of some provinces differ as widely from those of the others as a Chinese does from a Hottentot.'[18]

Tooke first addressed the problem of the expanse of inquiry. Information was needed on agriculture, diseases, industry, arts, trade, natural history, geography and meteorology. These should be grouped according to the 'several Russias' in the empire: 'Great Russia', 'Little Russia' (Ukraine), 'White Russia' (provinces of Poland) and 'New Russia' (Crimea). Not only was the amount of information unwieldy, but there were inconsistencies in transcriptions and translations of names. Clarke, who was orthographically conscious and went so far as to invent a font for printing his Greek transcriptions, protested against such careless practices:

> In the Russian alphabet there is no letter answering to our letter W; yet we write *Moscow*, and *Woronetz* [Voronezh]. Where custom has long sanctioned an abuse of this kind, the established mode seems preferable to any deviation which may wear the appearance of pedantry. The author has, in this respect, been guided by the authority and example of *Gibbon*; who affirms, that 'some words, notoriously corrupt, are fixed, and as it were naturalized, in the vulgar tongue. . . .' But, it may be fairly asked, where is the line to be drawn? What are the Russian names, which we are to consider *as fixed and naturalized in the vulgar tongue*? Are we to write Woronetz, or Voroneje; Wolga or Volga . . .?[19]

Travelling through the Russian provinces, Clarke corrected and added the names of provinces, towns and rivers, in standardised form on the maps he carried with him. 'On our maps [the river] is written *Donnez*; and in those of Germany, *Donetz*. We paid the greatest attention to the pronunciation of the natives; particularly of those Cossack officers who, by their education, were capable of determining the mode of orthography best suited to the manner in which the word is spoken; and always found it to be Danaetz.'[20] The river was accordingly renamed in Clarke's account.

Misrepresentations of the geography or nomenclature of the Russian provinces was not entirely the fault of the traveller. Ever suspicious of Tsar Paul's conduct, Clarke claimed that the government covered their dominions with a veil of secrecy, keeping foreigners ignorant of topography and obscuring traces of military conquest. Endeavouring to investigate the southern provinces, Clarke told of his encounters with the Russian police, who upheld the maxim that 'to enlighten is to betray':

> Had it not been for the jealousy of the *Russian* police, we might have published another more extensive view of the whole territory of the *Don Cossacks*; calculated to manifest the prevailing ignorance concerning the courses of the rivers, and the general geography of all the country bordering the *Sea of Azof*. It was prepared for us, in consequence of an order from the Governor

of the district, by a party of officers belonging to the *Cossack* army: but some agents of the police, apprized of the circumstance, endeavoured to excite a suspicion that we were spies, and we were not permitted to profit by their intended liberality.[21]

Clarke believed that efforts by the Russian police to disrupt their tour were rooted in a conspiracy to prevent foreigners from acquiring further knowledge of their imperial territory. Gaining detailed cartographic knowledge would not only mean accounting for barren spaces on western maps, but discovering the conditions of rule and the trajectory of Russian expansion. However paranoid Clarke may have been, he shared other travellers' determination to evaluate what territory was likely to be conquered next. Unsettled about the rise of Russia's power, and curious of the consequences this might have, many were anxious to chart its political manoeuvres.

Anxieties of expansion

'I cannot avoid beholding the Russian empire as the natural enemy of the more western parts of Europe,' commented 'Fum Hoam', one of Oliver Goldsmith's imaginary Chinese characters, while assessing the status of western civilisation. Russia was

an enemy already possessed of great strength, and, from the nature of the government, every day threatening to become more powerful ... The Russians are now at that period between refinement and barbarity, which seems most adopted to military achievement, and if once they happen to get a footing in the western parts of Europe, it is not the feeble efforts of the sons of effeminacy and dissention [*sic*], that can serve to remove them. The fertile and soft climate will ever be suficient inducements to draw which myriads from their native desarts [*sic*], trackless wild, or snowy mountain.[22]

Although a spoof on travellers' characterisations of civility on the Continent, for some contemporary reviewers Goldsmith's correspondents in *Citizen of the World* were considered to have held some reasonable views about Europe's political condition.[23] Hoam's judgement that half-civilised Russia posed a powerful threat to western parts of Europe was a sentiment shared by non-fictitious commentators. Attracted to its temperate climate, Russia might be able to get a foothold in the West unless Europe's comfortable, selfish and unmanly sons prepared properly. That Russia's conquests might capitalise on its idle and indulgent neighbours was a prospect in accord with other theories that climate and situation affect a nation's defences. Adam Ferguson, writing on the conditions upon which civil societies were built, spoke of European and Asian climates in relation to Russian conquests:

> The nations of Europe who would settle or conquer on the south or the north of their own happier climates, find little resistance: they extend their dominion at pleasure, and find no where a limit but in the ocean, and in the satiety of conquest. With few of the pangs and the struggles that precede the reduction of nations, mighty provinces have been successively annexed to the territory of Russia; and its sovereign, who accounts within his domain, entire tribes, with whom perhaps none of his emissaries have ever conversed, dispatched a few geometers to extend his empire, and thus to execute a project, in which the Romans were obliged to employ their consuls and their legions . . .[24]

The swift expansion of northern, eastern and southern borders impelled a number of British travellers to offer their own comments on the condition and aggression of the Russian empire. In 1772, Joseph Marshall, in an account of travels from St Petersburg through the Ukraine and the Polish borderlands, echoed others' warnings about what he sensed was the imminent demise of the Ottoman Empire, and believed that Prussia would have difficulties defending itself against further westward Russian expansion.[25] Nathaniel Wraxall, who toured northern capital cities, including St Petersburg in 1774, remarked that 'when one reflects on the immense magnitude of the [Russian] empire, one is lost in the idea.' It was growing too large for Europe's good. Russia seemed a power 'which we regard every day as more an object of political terror and watchfulness, and from whose arms Europe has even been taught to dread another universal monarchy.'[26] Another alarm was raised by the MP and Scottish agricultural 'improver' Sir John Sinclair, who, during a Parliamentary recess in late 1786, undertook a whirlwind tour of St Petersburg and Moscow, meeting the Tsarina Catherine along the way. He was critical of Russian government and suspicious of further aggrandisement. Considering the pace at which Russia was pushing southwards, he thought 'all Europe must unite to check the ambition of a sovereign who makes one conquest only a step to the acquisition of another.'[27] Having the capability to pressure the sovereignty of the northern states was another concern. A piece of political propaganda, attributed to Gustav III, sparked fears that Sweden might too become another Russian province. The tract, translated into English as *The Danger of the Political Balance of Europe*, was issued in 1790, the year of a peace treaty signed between Gustav and Catherine.[28] In 1788, taking advantage of the latest war between Russia and Turkey, Sweden attacked Russia, only to end up, two years later, settling on the same boundaries that had existed before the hostilities began. Following this Swedish failure, the tract stressed its chief enemy's aggression and record of hostile conquests. It characterised Russia as a European menace, disturbing the continental balance of power, ignoring the political turmoil of the French Revolution and warning other nations to monitor the outrages in the east. Catherine's was a 'vast empire, which, for twenty years, has spread terror, corruption, despotism, and war, embraces all varieties of climate, and comprehends every species of resource. Seas inaccessible to European fleets; deserts or enslaved countries are her frontiers.'[29] The principles of Catherine's court were despotic, and, unless checked by other continental states with a shared political philosophy, might corrupt and destroy the fabric of European civilisation.

Clarke concurred. While travelling through the southern Russian provinces a decade later, he cast his own judgements on the barbarity of the Russian conquests and the dangers of Russian expansion. His account of the sudden capture of the Crimea through a contrivance of Grigory Aleksandrovich Potemkin, the Overlord of Novorossisk and Azov and chief architect of Catherine's imperial policy, exposed a baneful plot to topple the ruling Tartar khan.

> The capture of the *Crimea* excited the attention of all *Europe*; but the circumstances which caused the desposition [*sic*] and death of the *Khan* are not so generally known. They have been artfully concealed by the *Russians*; and the brilliancy of the conquest of the *Crimea*, dazzling the imagination, has prevented a due inquiry into those dark and sinister maneuvers whereby the plot was perfected for the subjection of the Peninsula. *Potemkin*, archpriest of intrigue and wickedness, planned and executed the whole of it; to fulfil whose designs, it was immaterial what laws were violated, what principles trampled, what murders committed, or

what faith broken. His principal favourites were swindlers, adventurers, pimps, parasites: unprincipled men of every description, but especially unprincipled men of talent, found in him a ready patron.[30]

Potemkin promoted Russian advance in the peninsula, and pushed Catherine to annex the Crimea on the justification that a recent Turk incursion on Russian territory annulled a mutual treaty between Russia and Turkey. It was not so much the annexation of the Crimea that disturbed Clarke, however, but the alleged breach of diplomatic principles, ruthlessness and the burning passion to conquer.[31] Every Russian advance left a wake of destruction. Clarke complained of the difficulties he encountered in his attempt to investigate the ancient topography of the peninsula due to the prior arrival of the Russians.

> The ruins, as they still exist, with the assistance of . . . an accurate survey of the country, might be deemed sufficient for the purpose; but the insurmountable difficulties created by the barbarism of the *Russians* were very intimidating. When they settled in the country . . . [they demolished], and proceeding in their favourite employment of laying waste, they pulled down, broke, buried and destroyed every monument calculated to illustrate its former history; blowing up its antient foundations; tearing open tombs; . . . If the *Archipelago* should ever fall under the dominion of *Russia*, the fine remains of *Antient Greece* will be destroyed; *Athens* will be razed, and not a stone be left to mark where the city stood. *Turks* are men of taste and profound science in comparison with the *Russians*.[32]

We will have the opportunity to re-examine Clarke's opinions of the Turks when we examine the activities of British travellers to Greece in a later chapter. It is worth noting here, however, that he considered the path of destruction and the potential of southward Russian expansion a threat to the 'monuments' of antiquity and records of the history of modern Europe.

As he travelled southward, Europe was being swept with an imperial gusto from both sides: the Napoleonic march from France through Italy and Egypt, and Russia's encroachments from Poland to the Black Sea. Both Russian and French rulers compared their prowess to ancient imperialists, notably Alexander the Great. Controlling the ancient lands would be a highly symbolic conquest. But in the Ionian Sea, British forces were massing, preparing to check France's imperial progress and ostensibly rescue the ancient lands of liberty and civility from imminent conquest and barbarism. In these troubled times, travellers' accounts of Russian civility were particularly germane.

Peter's progress and Catherine's costly culture

If the beginning of the eighteenth century marked the beginning of Russia's rise as an imperial power in Europe, it also marked the beginning of a self-consciously motivated 'civilising process'. The extent to which its territorial expanses spread over Asia led to vast speculation about whether Russians were more European or Oriental in manners and culture. To be 'Asiatik' was, in European eyes, to be different – seen as the contrasting image of the European, and a subject of exploration and inspection in the emergent studies in 'Orientalism'.[33] The Orient was mysterious and laden with cultural stereotypes as being backward and uncivilised. The characterisation of Russia as Oriental reaffirmed its distinct identity. Lying somewhere between Europe and Russia was a gateway where a traveller exited European culture and civility and entered Oriental barbarism.

Peter the Great intended to change this perception, and looked to Western Europe for models upon which a new Russia could be built. From his initiative, travellers throughout the century observed the progress of Peter's civilising process. William Russell, writing a multi-volume *History of Modern Europe* at the end of the eighteenth century, commended Peter's achievements. 'Russia, under the auspices of Peter the Great, made a rapid progress in civilisation, and experienced perhaps the most sudden and fortunate change of any country of the same extent in the history of human affairs.'[34]

While Russia's conquests may have looked imposing to Europeans, Peter had designs to absorb and assimilate Western culture, and from his ascent to the throne in 1682, he set out to forge new relations with his European neighbours. In 1697–8, he embarked on a tour through Sweden, Prussia, Poland, Austria, Holland and England. His mission was partly diplomatic and partly intended to familiarise himself with European economic, cultural and intellectual life. Travelling incognito under the name of Sergeant Pyotr Mikhaylov, the Tsar took advantage of opportunities to work as an itinerant technician, learning artillery practices in Konigsberg and shipbuilding with the Dutch East India Company. He visited the

1 'Norr-Malm Square in Stockholm'. Royal Palace shown at the front and the Opera House to the left.

PORTRAIT of the REGICIDE ANKARSTRÖM
as he was exposed in the Streets of Stockholm.
during three days upon a Scaffold.

Published Jan.ʸ 1.1819. by T. Cadell & W. Davies. Strand. London.

2 'Portrait of the Regicide Ankarström' (tied to scaffold, assassin of Gustav III).

E. D. Clarke del.ᵗ   R. Pollard Sculp.ᵗ

MODE of EXHIBITING the BODIES of CRIMINALS in SWEDEN.

Published Jan.ᵉ 1 1819 by T. Cadell & W. Davies. Strand. London

3 'Mode of Exhibiting the Bodies of Criminals in Sweden' (engraving of a decapitated body on top of a tree trunk; the head and right hand are in separate trees).

View of the great MINE of COPPER at FAHLUN in DALECARLIA,

from a drawing by MARTIN of STOCKHOLM.

Published Feb. 1st 1822 by T. Cadell, Strand, London.

Letitia Byrne, sculp

4 'View of the great Mine of Copper at Fahlun in Dalecarlia'.

5 'Laplanders, having prepared their Winter Tents'.

6 'Mode of forcing a Passage through the Ice, when the Sea is not sufficiently Frozen to sustain the weight of the Human Body'. (Clarke and his fellow traveller Cripps off the coast of Finland, in 49 degrees Fahrenheit below zero weather, watching their guides break up the ice with their oars.)

7 'View of the interior of the Kremlin showing the ancient Palace of the Tsars and the first place of Christian Worship in Moscow'.

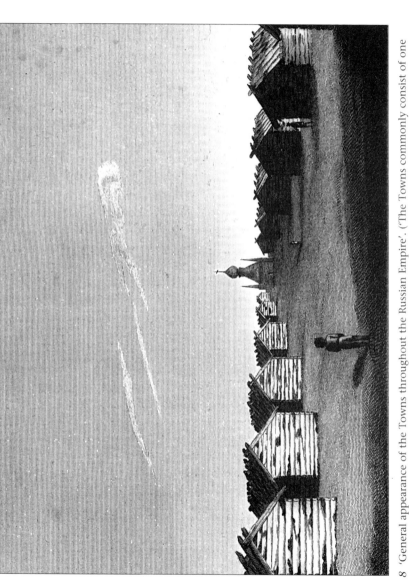

8 'General appearance of the Towns throughout the Russian Empire'. ('The Towns commonly consist of one street, as broad as Picadilly, formed by the gable ends of wooden huts, . . . and terminated by its church. The view of one of these Towns will afford the Reader with a very correct idea of the rest,' wrote Clarke (p. 34).)

9 'An Albanian of Greece'.

10 'An Athenian Lady in the Dress commonly worn'.

11 'Remains of a Temple of Jupiter Liberator at the base of Mount Gargarus . . . discovered by the Author in 1801'.

12 'The Pathenon at Athens' (the Parthenon being gradually dismantled by the French and British).

capitals, the hubs of intellectual life. In England, he attended a session of Parliament, visited the Royal Society, toured the Mint, and travelled to Oxford. He collected ideas for improvement and social reforms which, upon his return, he attempted to adapt to Russian culture. He set an ambitious programme and entered the eighteenth century with plans to consolidate Russia's strength, rise as a maritime power and organise a professional military. In sum, to bring a revived nation to a 'modern' European status.[35]

Peter's legacy

Educational reforms, institutional developments and alterations in the state's infrastructure were priority items for reform. Material progress was believed to be the prerequisite to moral improvement. Managing Russia's military might, size, diversity, industry and forging a newfangled high culture were acts governed by the ideology of a new 'secular state'. This emphasised a shift from ecclesiastical service and ecumenical ambitions of the Church, to new educational programmes for government administration, military maintenance and research in science and technology.[36] Lacking any previous platforms for education or training in jurisprudence, the sciences, or civil and military engineering meant that continental theories of philosophers such as Descartes, Leibniz, Pufendorf and Wolff were imported by recruiting foreign (mainly German) teachers. As a result, Petrine Russia has broadly been characterised as imperial over religious, cosmopolitan over parochial, and developing a governing modern intelligentsia over a ruling medieval Church.

These multifarious changes to Russia's social and political landscape were embodied in the creation of a new capital city, St Petersburg, whose foundation stone was laid by Peter in 1703. His European travels had impressed upon the Tsar an appetite for European style, which was reflected in the architecture of Petersburg. French and Italian architects designed buildings imitating designs found in London, Paris and Vienna. The Western influence was to modern Petersburg, with its wide, straight streets and square buildings, what the Byzantine influence was to medieval Moscow (see Plate 7).[37] What foreign travellers saw when visiting St Petersburg was a hodgepodge of familiar styles. Frantic construction meant that in 1712 Peter was able to move his court to the new capital, which represented not only the birth of a new cultural centre – symbolising the 'modernising' mentality of Peter's progressive reforms – but also stronger political and social affinity for European intercourse.[38]

Two early British residents in Russia under Peter the Great offered some preliminary perspectives on his society. John Perry, an ill-starred hydraulic engineer, met Peter during the Tsar's English tour and managed to secure a lucrative service as a technical adviser to Peter's various ambitious projects, one being to link the Don and Volga rivers, for £300 per annum. After fourteen years' service, Perry's most complete project was composing a catalogue of problems he had in dealing with the Muscovite nobility, the misconduct of the Russian admiralty and professional malcontent – a treatise which became the best known and perhaps most plagiarised British account of Petrine Russia in the eighteenth century. Despite his petition of grievances against the Russian government and military, he none the less gave some credit to the Tsar's attempts to improve his society, characterising him as a sort of dawning enlightened despot. Perry saw in the Tsar a fresh desire to release his people from the stranglehold of the Russian Orthodox Church and ecclesiastical conservatism by adopting Western culture, in style and reformed political machinery.[39] As William Tooke saw it at the end of the century, Peter saved no pains

to free his subjects by gentle degrees from the shackles of barbarism, to diffuse on all sides the benign light of arts and sciences, to discover the treasures discovered in his dominions, and to furnish agriculture with the remedies and assistances adapted to its improvement. His travels into several countries of Europe for the acquisition of such kinds of knowledge as were most applicable to the use of his dominions, are sufficiently known.[40]

Just as its long, broad, straight streets, paved with stone, were being laid down, St Petersburg required a population, and the nobility, Perry observed during his residence, seemed to Peter appropriate residents:

[A]mong some other causes, one of the chief which makes the generality of the nobility at present uneasy, is, that the Czar obliges them against their will, to come and live at Petersburgh, with their wives and their families, where they are oblig'd to build new houses for themselves, and where all manner of provisions are usually three or four times as dear, and forage for their horses, etc. at least six or eight times as dear as it is at Mosco.[41]

Peter not only selected the site, oversaw the careful planning of the new city and recruited its society, but also had his hand in regulating the customs of its population.[42] The city was to be Western through and through, as were its inhabitants, who were required to dress the part:

> It had been the manner of the Russes, like the Patriarchs of old, to wear long beards hanging down upon their bosoms, which they comb'd out with pride, and kept smooth and fine, without one hair to be diminish'd.... The Czar, therefore, to reform this foolish custom, and to make them look like other Europeans, ordered a tax to be laid, on all gentlemen, merchants, and others of his subjects (excepting the priests and common peasants, or slaves) that they should each of them pay a hundred rubles per annum, for the wearing of their beards, and that even the common people should pay a copeck at the entrance of the gates of any of the towns or cities of Russia.[43]

Not satisfied to stop at this prescription for making Russians 'look like other Europeans', Perry recalled how the Tsar also disliked the traditional long garments, pleated at the hips, which Russians were accustomed to wearing:

> The Czar therefore resolving to have this habit changed . . . gave orders that all his boyars and people whatsoever, that came near his court, and that were in his pay should . . . equip themselves with handsome cloathes made after the English fashion. . . . And next he commanded, that a pattern of cloathes in English fashion should be hung up at all the gates of the city of Mosco, and that publication should be made, that all persons (excepting the common peasants who brought goods and provisions into the city) should make their cloathes according to the said patterns.[44]

In addition to requiring his subjects to give up their kaftans and don Western attire, Peter also introduced the use of the Julian calendar to conform to European chronology, and encouraged Russian socialites to throw Western-style soirées and acquire a taste for coffee.

Some thirty years after Perry's residence in Russia, Jonas Hanway, a Portsmouth-born merchant of the Russia Company, was sent to St Petersburg and on to Persia to inspect trade relations, requiring him to live in and travel around the country for seven years, then under the rule of Tsarina Elizabeth. Hanway's *Historical Account of British Trade* was packed with views of the economy, navigation, climate and transport in Russia, as well as tales that made him into a swashbuckling adventurer dodging pirates, escaping enslavement, illness and rebel conflict. But his observations of the capital city bear testimony to the growing esteem shared by many of the plan of the city and its mix of foreign flavours. St Petersburg, he wrote in 1753, 'may at present be considered as the modern and polite metropolis, . . . an elegant and superb city'.

> It is divided by several canals, Peter the Great, intending to take Amsterdam as his model in building it; but from the reluctance with which it was originally begun by his subjects, who were compelled to build, and likewise from errors in the plan, some part of the city remains intirely unexecuted, and in others the houses are too near the canals. This does not hinder, but there are some regular, broad, and well built streets, and several very noble structures.[45]

While accounts of St Petersburg were limited in travel and historical literature throughout the eighteenth century in Britain, it eventually grew to be considered a cosmopolitan city with a diverse population, which, by Catherine the Great's reign, included a sizeable community of resident British merchants and traders. Tooke, while resident in St Petersburg from 1771 to 1792, calculated (conservatively) that 482 British resided in St Petersburg and Kronstadt in 1783. John Parkinson, an Oxford travelling tutor who travelled through Russia to Siberia between 1792 and 1794, put the figure as 1,500.[46] Andrew Swinton, a kinsman of Admiral Samuel Greig, a Scottish admiral in Catherine's service, lived in St Petersburg from 1788 to 1790. He described the capital as a growing international melting pot:

> I feel myself here as in another world, the dress, the manners, and customs of the people are so different from those of other nations in Europe. Besides the variety of nations which compose the Russian Empire, in my daily walk through the city I meet with English, Danes, French, Swedes, Italians, Spaniards, Portuguese, Venetians, Poles, Germans, Persians, and Turks; the latter are arrived here prisoners from Oczakow.

Like the empire, St Petersburg was a mixed bag which travellers found difficult to summarise, many accounts offering equally mixed views. 'Petersburg is a strange city,' summed up Swinton, 'even to the Russians: it increases daily, with new recruits from every corner of the empire.'[47]

The growing numbers of resident foreign merchants reflected the growth of Russian trade with Europe in the latter half of the century. The foreign population of merchants, taking advantage of increased trade privileges granted by the Russian government, seemed to some to trump Russians from their own trade. John Richard, who claimed to have toured Russia in 1775, reported (what he in fact probably learned from others) that 'notwithstanding the immense trade of Russia, it is remarkable there are few or no Russians who may be properly termed merchants'.[48] Samuel Bentham, who moved to Russia in 1780 to pursue lucrative investment opportunities, quickly assimilated himself to the 'English quarter' in St Petersburg, but was shocked to see how little his compatriots knew of Russia outside their small community. The resident British, he said, 'despise everything Russian, and know no more of the country or people than what they read in the English newspapers of the Petersburg intelligence.'[49] In 1786, encouraged by his brother's promises of financial gain, Jeremy arrived in St Petersburg and was similarly appalled at the catchpenny accounts of Russia he had received in Britain. Writing to his father shortly after his arrival he advised, 'whenever you see in a newspaper an article from Petersburg lay 2 to one . . . that it is not true.'[50]

Ironically, the ignorance that prevailed about Russia seems symptomatic of the very existence of the resident English community in the capital, whose penchant for insularity encouraged them to disregard their surroundings. Even with the diversity of people living within the Russian territories and cities, and of Europeans who lived in and influenced the building of the new nation, astoundingly few accounts provided an accurate portrait of the rich cultural diversity which permeated society. Many misleading statements were plagiarised, unwittingly propagating prejudiced views, which did not improve British impressions or ignorance. Due to such circumstances, outdated and chauvinist characterisations of Russian culture were widespread. A common message sent home by entrepreneurs and merchants was that commercial opportunities prevailed for the British because the Russians, despite their Western façade, lacked the knowledge, skill and civility necessary for commercial success.

'The truth is,' explained Swinton, 'the Russians are going too fast in affecting, as well as attaining, improvement. Foreigners have put too many things into their heads, and, I believe, are picking their pockets, by the idle schemes, with which they amuse them. The Russians, in general, look upon foreigners as a kind of superior beings, in regard to the arts and sciences.'[51]

The 'westernising' process, which for foreign visitors to Russia equated with a 'civilising' process, was rapidly designed and imposed upon Russian culture by Peter I at great expense but with little foundation. The physical difficulties in building St Petersburg – a city built on marshlands at a cost of 30,000 lives – was analogous to the problems involved in acclimatising Russians to Peter's ideological dreams. The framework for a new Russia was built with the expectation that people would grow into it. His visions overlooked the lack of available resources. This was exemplified in his ambition to build an Academy of Sciences akin to the Royal Society of London or Académie des Sciences in Paris, which materialised less than a year after his death, which neglected the fact that (due to lack of a university) no native talent could staff it.[52]

Some British writers pointed to these anomalies in Peter's plans for the development of a modern Russian state as proof that the Russians remained an inferior race, ruled by a tyrannical government. Others, like Swinton, objected:

> Some will say that the Russian nation are not yet civilised; and that Peter only began the work of civilisation – of arts and sciences. What narrow thought! – ... He gave the plans of the building – he laid the foundations, and reared a part of the walls: succeeding Monarchs are his workmen, his bricklayers, slaters, carpenters, painters, and upholsters. Catherine II is the most distinguished of Peter's work people, and has made such elegant improvements upon the original plan, that it is so far become her own.[53]

Peter's reign was seen as the foundation of the modern Russian empire, and for most of the eighteenth century travellers placed their observations of institutional and social developments within an historical context of Peter's original plans for reform. As William Coxe remarked in 1784, 'Voltaire's *Life of Peter the Great* ... is the work from which most foreign nations have formed their ideas of Russia' (a work, it could be noted, that was commissioned by Peter's agents!).[54] But as more sympathetic writers, such as Swinton, suggested, Peter's reign should be viewed as the beginning and not the result of the civilising process. It was disingenuous to criticise initial difficulties in reforming a vast society; travellers should judge the nation by her sons rather than the fathers.

By Catherine's reign, commentators had a growing cultural panorama to consider in their critiques of the condition of modern Russia. St Petersburg gained the Academy of Sciences (1725), the Cadet Corps (1731), Academy of Fine Arts (1757), the Institute of Mining (1773) and, in Moscow, the first teaching hospital (1707), Moscow University (1755), and the Russia Academy (1783). While new pursuits in the fine arts and sciences continued to bud between the death of Peter the Great (1725) and Catherine the Great's accession to the throne (1762), this period is still often dealt with as one of relative decadence, in favour of the epochs of the 'Greats'. This tendency began in the late eighteenth century, when the 'improvements upon the original plan' by Catherine during her 34-year reign became representative of a second epoch of reform. How did she fare? Did she finish wrapping Russia for presentation as a modern, civil state?

Catherine's legacy

Other than continuing educational and political reforms, Catherine's reign has become notable for building an elegant court and intellectual culture. But in the eighteenth century, the merits of her achievements in this area – in her choices and initiatives to extend a 'European enlightenment' to Russia – were reviewed differently by various British travellers.[55] By the time of her reign, British travellers entered Russia anticipating an encounter with the convergence of 'Asiatic pomp and Eastern magnificence' with European philosophy.[56] Catherine, born in Germany an impecunious princess, acquired the status of Empress of Russia through a coup d'état against her husband, Karl Ulrich (Peter III), just months after he ascended the throne.[57] Known to be ambitious, energetic and a strong ruler, but also profligate, promiscuous and (in some views) personally unstable, many foreign spectators oscillated between admiration and obloquy. She gained early favour among many of the Russian nobles for her liberal views and fondness for Enlightenment philosophy. From the start of her reign, she set out to create a court to rival Versailles, befriending a number of French luminaries and resolving to establish new laws and foreign social relations.[58]

Eager to gather knowledge about the resources and present condition of her empire, Catherine launched a wide-ranging programme to collect information from all territories, recruiting foreign naturalists and scholars through the auspices of the Academy of Sciences as her explorers. Among them were the travellers referred to previously: Georgi, who studied the flora and fauna of the Urals; Gmelin, who explored Persia; and Pallas, who studied the natural history of Siberia. By the end of the 1770s, these travellers had trekked many miles, but systematic analysis of the data was still lacking and few works were available in Britain. Tooke, summarising these travellers' accounts in his *View of the Russian Empire*, lamented the time it was taking for these travellers' collections to be properly ordered.

> Very few of the accounts that have been given by travellers contain so great a variety of new and important matters. The journals of these celebrated scholars even furnish such a great quantity of materials entirely new, for the history of the three kingdoms of nature, for the history of the earth, for rural oeconomy, in short, for so many different objects relative to the arts and sciences, that it would require . . . whole years and the labour of several literary men only to put these materials into order, and properly to class them.[59]

At risk of redundancy, Catherine continued to recruit travellers and surveyors, encouraging more rapid publication of their results. But here her literary endeavours only began.[60]

Catherine's patronage stimulated literary interests, and a new generation of Russian writers, publishers and translators was born. Metropolitan Russia was charged with new intellectual currents, waking educated Russians and nobles to a new sense of cultural identity.[61] Many aristocrats assembled private libraries and avidly collected books on etiquette, heraldry and the latest works of the *philosophes* (Catherine even bought Diderot's and Voltaire's book collections which she added to the imperial library).[62] Catherine, who since childhood held high regard for French intellectual culture, saw salons, social clubs and banqueting halls spawn throughout Petersburg. French became the language of the Russian court, and French tutors were hired to educate patrician children.[63]

Sir James Harris, the British envoy and literary critic of some note, drew up an 'Account of Literature in Russia, and of its Progress towards being Civilized', published in England in 1780. His list was a selective account of classical literature that appeared in St Petersburg, in translation and at use at Moscow University. Shortly after, William Coxe's account of the university's collection of Greek manuscripts, as well as a lengthy survey of the state of literary endeavours in Russia, appeared which puffed up the view that Russians were on the right road towards being civilised.[64] For Western reviewers of the literary scene, especially those educated at Oxbridge, classical literature was the point of departure for progress along that road. Having translations of Latin and Greek texts in the library represented a commitment to sharing the classical heritage and Western values.[65]

By the 1780s, the major classics in European literature were available, and Catherine was able to read Shakespeare in German translation (though she preferred Sterne's *Tristram Shandy*).[66] Many books were from the production line of the Society for the Translation of Foreign Books into Russian, created by Catherine in 1768.[67] But some travellers, such as Clarke, were not convinced that the appearance of Western fiction would help the civilising process. In particular, Clarke found that English romance does not pass through cultural filters:

> Of the [English novels], the 'Italian' of *Mrs. Radcliffe* has been better done than any other; because, representing customs which are not absolutely local, it admits of easier transition into any other European tongue. But when any attempt is made to translate *Tom Jones, The Vicar of Wakefield*, or any of our inimitable original pictures of English manners, the effect is ridiculous beyond description. *Squire Western* [sic] becomes a *French Philosopher*, and Goldsmith's *Primrose* a *Fleur de Lis*.[68]

Non-fiction works were also criticised by Clarke to the extent that he concluded that books 'of real literary reputation are not to be obtained either in the shops of Petersburg or Moscow'. What they sold were 'gaudy French editions' of the science populariser Bernard de Fontenelle and the novelist Jean François Marmontel, Italian sonneteers and English folios of butterflies, shells and flowers. In short, 'the toys, rather than the instruments of science'.[69] As far as intellectual stimulation and original literary production, the Russians were far behind their European models. Clarke remained unconvinced of Russia's road to Enlightenment, seeing in Catherine mimicry, not creativity. 'In whatsoever country we seek for original genius, we must go to Russia for the talent of imitation. This is the acme of Russian intellect, the principle of all Russian attainments. The Russians have nothing of their own; but it is not their fault if they have not every thing that others invent.'[70] For others more favourable to Catherine's efforts, such a criticism would have been seen as too unsympathetic of the difficulties in maturing and improving society. After all, 'Civilization is best promoted by example,' opined Swinton, whose book was dedicated to Catherine II. What's more, the 'British nation is copied by all others, because they are the richest.'[71] Clarke's critique was partial, and, as further discussed below, motivated by anger rather than by critical assessment. In fact, Catherine's literary interests extended beyond publications on the natural history of the Russian provinces, which Clarke chose to forget, to the work of the newly founded Russian Academy and the compilation of the first dictionary of the Russian language, which was issued in six volumes between 1789 and 1799.

Nor were the arts neglected during her reign. The Academy of Arts, founded just before Catherine's *coup*, was the beneficiary of heavy endowment by the empress, concerned to collect classical relics and models. In the hall of the Academy stood a cast model of the Apollo Belvedere, and the artists and architects trained there were responsible for the construction of palaces, public buildings, villas and private residences in the classical style. The English painter and prolific travel writer, John Carr (referred to by Byron as 'Europe's wandering star'), who toured Petersburg in 1804, absorbed the delights of Catherine's 'profuse magnificence'. 'Russia is unquestionably much indebted to the genius and spirit of the late Empress; but it was impossible that *extended* civilization could be the fruits of her costly culture.'[72] In generously supporting the arts and sciences, Russian culture won magnificent monuments that continue to honour Catherine's reign. But her celebrated legacy was limited to the metropolis. High culture was purchased at the cost of the neglect to improve the life of the peasants in provinces.

Carr pointed out the limits and shortcomings of the enlightened despot's reign, and, shifting attention from court culture to peasant poverty, exposed Catherine's short-sighted policies in regard to the serfs. Her *Great Instruction* (1767), a manifesto inspired by her reading of Montesquieu's *Esprit des Lois*, the *Encyclopédie* and Cesare Bellaria's *Delle delitte e delle pene*, which proposed reforms to the constitution and a new code of laws, was a comprehensive expression of her political ideology. Humanitarian issues in the *Instructions* generally favoured religious toleration and disapproved of capital punishment, torture and the perpetuation of serfdom, but, regarding the last point, opposed any general measure of emancipation. While these may have represented Catherine's views, the *Instructions* seems to have been opposed by members of the nobility, who feared that they would lose power and authority over the serfs.[73] In the end, continuing debates over the practical reorganisation of the government stagnated and eventually killed off the *Instructions*: its principles were not put into effect and, by the end of her reign, the plight of the peasants had actually worsened. Serfdom increased (being imposed in the Ukraine) and serfs lost the small remnant of liberty they had and could become slaves – the personal property of their owners.

John Williams saw the failure of Catherine's reforms over the issue of serfdom as a mark of failure in the civilising process. Catherine's commitment to securing her absolute authority pointed to the problems and insecurities of a despotic society, and her failure to abolish slavery revealed the remnants of the 'wickedness of their ancient oppressors'. In the 1770s, Williams recognised that social improvement still had some way to go. 'Notwithstanding the great improvements that have been made towards civilising the Russian empire,' he wrote, 'the nobility and gentry of this nation, except those who are immediately about the court, are still brutal and tyrannical to a great degree, to which the arbitrary power which they have over their slaves not a little contributes.'[74]

A few years later, the Scottish tutor William Richardson, also secretary to Lord Cathcart (Ambassador Extraordinary to the Court of Catherine II between 1768 and 1772), characterised Russian class relations as 'the supreme authority of the Emperor over the nobility, and of the nobility over their slaves'.[75] Richardson, contemptuous of despotism and slavery, was nevertheless hopeful of the future liberation of the Russian servile system, and believed that their present condition and social hierarchy were in a transitional stage, finally moving beyond centuries of ancient feudalism. Britain was exemplary. He held hope by linking historical forces active in Russia with Britain's own providential history, where the latter's advanced civilised status created a positive model for the trajectory of Russia's future. 'There is some satisfaction in recollecting', he wrote, 'that while other nations grown under the yoke of bondage, the natives of our happy islands enjoy more real freedom than any nation that does now, or ever did, exist.'[76] Even Carr, who admired

Catherine for promoting and patronising the fine arts, thought her reign would have been more triumphant had she cured the evils of serfdom: 'As far as my observation and information extended, I should conceive that the civilization of Russia would be rapidly promoted, after the removal of that most frightful and powerful of all checks, slavery, by improving the farms...'[77]

Tensions over this issue increased in 1790 when a Russian nobleman privately published his own bitter attack on serfdom and autocracy. Alexander Radishchev's *Journey from St Petersburg to Moscow* described the dehumanisation of the serfs and the corruption of their masters, warning that these conditions would continue to tear away at the Russian social fabric.[78] Incensed by the book, Catherine had Radishchev arrested and condemned to death, but the sentence was commuted to exile in Siberia. Radishchev's timing did not help his cause. In 1773, Emel'ian Pugachev, a Cossack from the Urals, proclaimed he was Peter III (Catherine's assassinated husband) saved from the dead, and led the provincial peasants into a fierce rebellion with the aim to overthrow and annihilate the hated nobility.[79] That rebellion was quashed and Pugachev beheaded in 1775, but Catherine's ardour for libertarian reform was further cooled by the outbreak of the French Revolution in 1789, an event which sent chills up the spines of monarchs throughout Europe. The publication of Radishchev's attack on her rule, topped with the events of 1789, drove Catherine to tighten the bonds on her people, representing a switch from leading a modernising to a restraining political regime. Catherine's treatment of the problem of serfdom was as ignominious as her patronage of the arts was triumphant.

The execution of Louis XVI in 1793 made matters more sensitive. Travellers during this time drew particular attention to comparative European political rule and the crises of national pride and social identity that it created in different nations. Governments, Carr noted, had to be quick to act, as an anecdote of his revealed. Catherine put down a restless sect and 'saved her people from the baneful contagion of French principles'.

> During that revolution, which portended ruin to all the sacred establishments of all nations, when in England Pitt trampled out the brightening embers, and saved his country from the devouring flames, a group of mischievous emissaries from France arrived at Petersburg, and began, in whispers amongst the mob, to per-

suade the poor droshka driver, and the ambulatory vendor of honey quass, that thrones were only to be considered as stools, and that they had as much right to sit upon them as their empress: Catherine, concealing her real apprehension, availed herself to the powers with which she was clothed, without shedding a drop of blood.[80]

While managing to quell domestic discontent and generally saving prestige as an 'enlightened' despot, political turmoil none the less created rough seas for sailing into the nineteenth century. The last decade of Catherine's reign was marred by the menace of the French Revolution, Jacobinism, peasant rebellions and increasing constrictions on her subjects' liberties. For her son and successor to the throne, Paul I, these problems plagued his entire, albeit short-lived, reign (1796–1801).

Paul: 'The most barbarian among Christians'

Tsar Paul I opened his reign disliking his mother, distrusting foreigners (albeit most prominently Jacobins) and insecure about his own authority as ruler of Russia.[81] He began his reign reasserting the principle of absolute autocracy, set about reversing the policies previously established by Catherine, banned foreign books and music, withdrew permission for his subjects to leave the country for travel or study, and forbade French visitors to enter without a passport signed by the Bourbon princes to prove they were anti-revolutionary.[82]

After Paul's accession in 1796, fears over France grew with every advance that General Napoleon Bonaparte made in the eastern Mediterranean. In 1798, concerned about Napoleon's recent occupation of Egypt, Paul joined the Second Coalition, which ultimately included Austria, England, the Kingdom of Naples, Turkey and Russia.[83] However, for various reasons, Paul angrily withdrew from the coalition. The first was Russia's troubled relations with Austria (partly brought to a head by Paul's ambivalence about, and Austria's strong interests in, a military campaign in Italy). Second, Paul was incensed over a number of failed military advances made in collaboration with England (including an Anglo-Russian attack on Holland in 1799). He was also increasingly suspicious of increased British presence in the Mediterranean.[84] Following his withdrawal, he ventured into peace negotiations with Napoleon, going so far as to launch an expedition against English possessions in India, towards which end over 20,000 Cossacks were dispatched. However, these anti-British moves had disastrous consequences for the emperor, and in March 1801, just months after deploying his troops, high-ranking Russian conspirators, loyal to Catherine's principles, assassinated him.[85]

The five years of Paul's reign, then, created unpleasant circumstances for British travellers. The most outspoken critic of Russia, as it was experienced in 1800, at the height of Paul's anti-British policies, was Clarke. Because of what he called the 'madness and malevolence of a suspicious tyrant', he found much to condemn in Russian society and manners, whether 'Prince or peasant'. He described the villainy of the police, public reprimands, physical punishment, bribery and murder as standard elements of Russian life. English visitors walked through the streets of the capital at high risk, feeling like prisoners on parole. All Russians were servile to superiors. Downward from the throne, all were dependent upon the breast of the body above them for their livelihood, and all superiors abused their powers. 'The Emperor canes the first of his grandees; princes and nobles cane their slaves; and the slaves, their wives and daughters. Ere the sun dawns on Russia, flagellation begins; and throughout its vast empire, cudgels are going, in every department of its population, from morning until night.'[86]

Far from pursuing a path to enlightenment, Clarke believed, Paul routinely condemned creativity, disrupting a promising course of improvement set by his predecessors. 'Since the death of Catherine, it seemed to be the wretched policy of their Government to throw every obstacle in the way of intellectual improvement. Genius became a curse to its possessor; wit, a passport to *Siberia*.' Due to apathy and ignorance, the desire to pursue truth and science were left in abeyance, leaving the condition of society stuck in the past. Unlike the eulogistic characterisations of past Russian rulers offered in the *Modern Universal History*, a set reference text for modern history lectures at Cambridge, the Russians were not nearer to reaching modernity.

> The more we enquire into the real history of *Russia*, and of the Russian *Sovereigns*, the more we shall have reason to believe, that the country and people have undergone little variation since the foundation of the empire. PETER THE GREAT might cut off the beards of the nobles, and substitute *European* habits for *Asiatic* robes; but the inward man is still the same. A Russian of the nine-

teenth century possesses all the servile propensities, the barbarity of manners, the cruelty, the hypocrisy, and the profligacy, which characterises his ancestors in the ninth.[87]

Time did not mend Clarke's impressions of the society or his perception of the treatment of the English. Surviving Russia was akin to surviving the wild 'dark continent'. 'Mungo Park could hardly have been exposed to a more insulting tyranny among the Moors in *Africa*, than Englishmen experienced at that time in Russia, and particularly in *Petersburg*.' He had little wonder about the fate of the emperor shortly after his departure. 'Viewing the career of such men, who, like a whirlwind, mark their progress through the ages in which they live by a track of desolation, can we wonder at the stories we read of regicides?'[88]

Clarke's acrimonious account upset some fellow travellers and reviewers who thought he was over-prejudiced and unfair to broader historical improvements in Russian society. But, declared Clarke, had he not the right, indeed the duty, to present as vivid account of his experiences as possible? Unlike the French *philosophes* whom Catherine persuaded to prostitute their venal pens in order to 'varnish the deformities of her reign and empire', Britain would, he hoped, 'forgive the frankness of one, among her sons, who had ventured, although bluntly, to speak the truth.'[89] Some were forgiving, but thought that criticism against a country with whom England had been allied against France and who was generally sympathetic to eliminating the contagion of French tyranny might appear politically insensitive.

Henry Brougham, in his favourable review of Clarke's first volume in the liberal *Edinburgh Review*, immediately attempted to diffuse possible misconceptions of Clarke's attack on Russian society by warning that Clarke 'may be mistaken for a *Jacobin* . . . or perhaps persecuted as a *Papist* – or, peradventure, described as favourable to *French Principles*'; however, the reader's consciousness should be put at ease when reassured that 'there is not a word about English politics in Dr. Clarke's work; . . . it gives us a plain report of what the author did, saw and heard.'[90] Clarke was not criticising Britain's alliance with Russia, nor attacking Russian conduct as far as their attempts to defeat the French. But due to different political prejudices, travel authors ran high risks of having their motives for writing critical critiques of European governments. In the years Clarke represented as gloomy and repressive, the ultra-Tory *Anti-Jacobin*

Review was praising Paul's 'spirit and magnaminity'. Far from barbaric, the emperor 'has displayed the strongest marks of political wisdom with the present war.' In 1800 – the year Clarke judged him mad and malevolent – the *Review* predicted Paul would prove 'the SAVIOUR OF EUROPE'.[91] Even the year after publication of Clarke's book, the journal ran a commentary expressing British approval with Russian foreign policy towards France.

Brougham was right to point out possible problems, but, as Clarke explained in a note to his cousin, 'Never let political intrigue or party prejudices persuade you that I have *"ought extended or set down in Malice"* concerning the Russian people.' He believed his account was controversial simply because 'in England it is not *quite convenient*, in the *Cabinet*, to have it thought that the Russians are . . . THE MOST BARBARIAN OF CHRISTIANS.'[92] Brougham had reason to be sympathetic. On his own northern tour of Scotland and Scandinavia in 1800, Brougham decided against extending the trip to St. Petersburg, citing the 'extremely disagreeable situation of travellers residing in Russia . . . Contrary to my expectations – I found despotism the more hideous the nearer I attempted to approach it.'[93]

Other British travellers during Paul's reign who were cautious about crossing into Russian territory were not as outspoken as Clarke about their feelings. Even Malthus, on his way from Stockholm to St Petersburg, worried about being allowed to enter Russia at all, 'as the Emperor is so extraordinarily fearful of admitting strangers into his dominions. We hear that the last Englishman that went from Stockholm was allowed with some difficulty to pass the frontiers, but his servant and baggage were sent back and he was only allowed to take a couple of shirts with him.'[94] Hearsay spread amongst travellers, as well as rumours, prejudices and blatant misrepresentations of foreign lands. But waging vehement accusations about the barbarity of Russian society based on such a narrow window of history, and such limited experience, was thought by others to simply deny justice to broader cultural developments, and did not reflect the progress made since Peter's reign.

Matthew Guthrie, the Scottish physician to the Court of St Petersburg, wrote that he and his wife Maria Guthrie 'have lived 37 years in Russia, and have had no reason to consider the inhabitants worse than those of other countries, nay even the courtiers we regard as much the same as the favoured sons of fortune in other climes; there is nothing so unphilosophical as those attempts to vilify a whole nation.'[95] The British military attaché Sir Robert Wilson, himself known to have written scathing remarks on the conduct of the Russian army, wrote to Clarke objecting to his sweeping condemnation of Russian society.[96] Clarke confessed that travellers at different times – those writers who visited Catherine's Russia or later travellers (like Wilson) to Alexander's Russia, might have different views, but defended his public abuse of the late Tsar. Even in 1823 travellers persisted to write lengthy objections to Clarke, such as Robert Lyall's defence of *The Character of the Russians*.[97]

Travellers debated the accuracy, intentions and prejudices of different accounts, each proffering their own as more authoritative than others'. Over decades, travellers peeped through small windows of Russian society, and variegated views create a spectrum of opinions, castigation and fond reflection. In the end, most determined that despite the reform movements spun into motion by Peter and reinvigorated by Catherine, the system of civil laws at the end of the eighteenth century remained imperfect, and ill calculated to provide justice and order in the country. John Williams offered a reasonable assessment in his account of Russia mid-way through Catherine's reign:

> Upon the whole, I think we may justly observe, that till the inhabitants of Russia are restored to that state of liberty from which they have been most iniquitously driven, and till they are suffered to enjoy the natural rights of mankind, and to think and act like rational beings, this great empire will never be famous for her arts, manufactures, or commerce; and that till this firm foundation is laid, whoever attempts by splendid codes of laws, or by any other means, to effect this great work, will be acting like the man in the gospel who built his house upon the sand, which was overturned by the natural efforts of the elements, because it was unnaturally formed.[98]

Like the early concerns over the uncertain foundations of Peter's new capital city which provided a 'window on the West' ('built on tears and corpses,' remarked the Russian historian Nikolai Karamzin), it was clear that the stability and progress of a new Russian body politic needed a firm ideological foundation.[99] But the capital city was merely one lens through which Russian politics could be viewed, and its residents only representatives of the 'body' of Russia. In the age of colonisation and imperial expansion, the condition of the peasants provided as relevant information towards defining the civility of Russia as the philosophy of the Princes. While living in the provinces, the varied populations of Cossack, Caucasus and Kalmyk tribes were central to the political and historical debates on the 'civilising process.'

Warriors in the 'unknown territory'

After only a short, but unpleasant, time in St Petersburg, Clarke and Cripps headed south. On 1 June they left Moscow, 'where we passed some pleasant hours, and many others of painful anxiety, exposed to insult, and to oppression, from the creatures, spies, and agents, of the contemptible tyrant who was then upon the Russian throne.' Their plan was to visit the Crimea by a circuitous route, through the territory of the Don Cossacks and the more distant regions of the 'Kuban Tartary' and 'Circassia'.[100]

Clarke and Cripps sailed down the Don making notes on the manufactures, trade and natural history. They travelled through Voronezh (where merchants brought furs from Siberia and silk, porcelain and precious stones from China), Pavlovsk (where he commented on the trade of tallow and fruit) and towards the territory of the Cossacks. Since it was a part of the country 'rarely visited', Clarke spent much time discussing the 'independent mode of life of the people; their indolence at home; their activity in war; their remote situation with regard to the rest of Europe; the rank they hold in the great scale of society; the history of their origin; their domestic manners, and character.'[101] And, not least, in appearance it was a stark contrast to the metropolis (see Plate 8).

Cossacks and 'Calmucks'

In Kazanskaya, on the river Don, north-west of what is now Volgograd (Stalingrad), they first encountered a community of Cossacks. Clarke found their appearance dignified and majestic, but also intimidating: 'his elevated brows, and dark mustachoes [sic]; his tall helmet of dark wool, terminated by a crimson sack, with its plume, laced festoon, and white cockade; his upright posture; the ease and elegance of his gait; give him an air of great importance.'[102] The Cossacks were peasant-warrior tribes scattered about the hinterlands of the Black and Caspian Seas. They maintained self-governing military and territorial units. With the empire's southward expansion throughout the eighteenth century, the Russians gained increasing control over Cossack administration. In return for military services – as defenders of the Russian frontier and advance guards for the territorial extension of the empire – the Cossacks managed to restore some of their autonomy. By the nineteenth century, however, the Russian government had abolished a number of senior Cossack administrative positions and increasingly came to dominate the tribes.[103]

The Cossacks – whose name has a Turkish etymology meaning 'wanderer' or 'free man' – did not idly accept usurpation. Without war they seemed lost. 'A quiet life seems quite unsuited to their disposition: they loiter about, having then no employment to interest them; and being devoted to war, seem distressed by the indolence of peace.'[104] Their warrior reputation was inspiring, so that even the English poet Thomas Tickell, in his 'on the prospect of peace' in Europe (1713), would write:

> Let grim Bellona haunt the lawless plains
> Where Tartar clans and grizly Cossacks reign.[105]

But they were vilified by the Russians, who throughout the eighteenth century incessantly fought to quash Cossack revolts against their authority. Russian officials, Clarke believed, propagated lies and led foreigners to fear the perils of exploring the remote territories. 'In *Russia*, there was not an individual, of any respectability, with whom we conversed upon the subject of our journey, who did not endeavour to dissuade us from the danger of traversing what was termed "the deserts of the Don Cossacks".' He found these assertions absurd, and suggested that, on the contrary, it was amongst those in the capital that 'we were constantly exposed to danger', but amongst the Cossack warriors they faced 'a brave, generous, and hospitable people'.[106]

Clarke created contrasts between the metropolis and the provinces; the residences of the princes and dwellings of the peasants, inverting expectations of where 'barbarity' was to be found. 'In times of hostility the *Russians* found in the *Cossacks* a desperate and dangerous enemy; and many a bitter remembrance of chastisement and defeat induces them to vilify a people whom they fear.' Wilful misrepresentations by the Russians were part of the apparatus of creating dominion over and sanctions against their southern subjects. Revolts and unrest were to be expected. 'The *Cossacks* are therefore justified in acting towards them as they have uniformly done; that is to say, in withdrawing as much as possible from all communion with men whose association might corrupt, but could never promote, the welfare of their society.'[107]

Not prepared to share the wealth of her empire, the administration under Catherine II waged a number of assaults against the self-governing Cossack enclaves to break down the notion of a separate identity for 'Little Russia'. But the Cossacks resisted any attempts at assimilation or systematic colonisation through revolts. The Pugachev uprising was the most notorious of many other movements in which they participated. But territorial tensions did not end with the contests between the Cossacks and the Russians. In the southern frontier lived a number of different peoples. The steppes were populated with extensive systems of tribal kinship rather than mixed nationalities cohabiting an area. Semi-independent Cossack tribes also negotiated military services for the Poles and fought against the Tartars, Turks and Muscovites. In an attempt to draw a reasonable picture of the southern frontier, and make sense of the complicated networks of trade, territorial occupation or relative condition – savage or civilised – of the inhabitants, Clarke took time to traverse the region.

Escorted by Cossack guards who 'scoured the plains, armed with pistols, sabres, and lances twelve feet high', Clarke and his party set out to encounter and describe other tribes in different territories. In the dry mid-June sun (averaging in the eighties Fahrenheit according to Clarke's daily log), they ventured west, over the steppes in what is now the eastern Ukraine, looping from 'Kamenskaia' [Kamensk] around to 'Tcherkask' [Novocherkassk]. In this flowery wilderness they found a camp of 'Calmucks' [Kalmyks], who were startled and confused over the approaching caravan with armed guards galloping toward them. As they drew near, 'about half-a-dozen gigantic figures came towards us, stark naked, excepting a cloth bound about the waist, with greasy, shining, almost black skins, and black hair braided into a long queue behind.'[108] Being 'a little intimidated' at these screaming, naked bodies running towards them (nakedness alone being enough to offend the sensibilities of an English 'literary traveller', and being the most cited characteristic of the savage from Cook to Darwin), Clarke plucked up the courage to step forward with his hand extended, 'which seemed to pacify them'. Upon closer inspection of their living conditions, he noted the physical similarities between the Laplanders and the Calmucks, appearing to confirm the eastern origin of both, despite the former being 'dwarfs' and the latter 'giants'. Despite the sheer distance separating these groups, similar customs, languages, tents and even brandy seemed to support further his conclusion. 'We are not otherwise authorised in comparing together tribes so remote with each other, than by asserting from our own observation, that both are *Oriental*, and that both are characterised by some habits and appearances in common.'[109]

The Calmucks were a people for the countryside, and were resilient against the interference from outside their community. Like the Cossacks, when they were not waging war, they pursued entertainment rather than 'improvement'. They enjoyed hunting, wrestling, archery and horseracing; in winter, they played cards, draughts, backgammon and chess. They spent entire nights gambling and playing two-stringed lyres. 'In short,' Clarke concluded, 'it may be said of the Calmucks, that the greatest part of their life is spent in amusement.' And why not? These people on the fringe, occupying the hinterlands of civilisation, equipped themselves with enough to survive and enjoyed their recreation. They lived and loved the outdoors. Both Gmelin and Pallas had previously described the Calmucks' repugnance for living in closed quarters, and the horror with which they enter towns when necessary. They resisted cultural assimilation and conforming to westernised ideals of civil living. Surrendering any evangelical urge to encourage improvement and education in savage communities, Clarke thought it best that they be left alone. 'Wretched and revolting as it may seem, they would be indeed miserable, if compelled to change their mode of living for that of a more civilized people.'[110]

Clarke donned his zoologist's hat and ran through a catalogue of the animals found in the provinces, which he compared to Shaw's *General Zoology*, and he continued his cartographic project.[111] Despite difficulties in eluding the censorship of the Russian police, his maps, filled with revised place-names and the delineation of different rivers, were offered to the 'geographers of Europe'. 'Those steppes which are described as being so desolate, and which appear like a vast geographical blank in every atlas, are filled with inhabitants.'[112] By the end of June, they arrived in the Cossack settlement of 'Åxay' [Aksay]. Mistaken for an English general, Clarke's arrival at Åxay was treated like a royal entry festival, with Cossack cavalry, officers and the 'Ataman' (a sub-leader of the military district) present to greet their visitor.

The festive spirit flourished the following day when the town celebrated the recovery of one of the Emperor's children from a smallpox inoculation. A Cossack general and commander-in-chief of the district invited Clarke and Cripps to join his staff officers for a dinner and a public ceremony in the church, which involved singing, praying and reading from the Psalms, but by no means privation of eating and drinking. The table was covered with 'all sorts of fish, with tureens of sterlet soup, with the rich wines of the Don, and with copious goblets of delicious hydromel or mead, flavoured by juices of different fruits'.[113] Later, again banqueting with the same Cossack general, with the women 'amusing themselves with a piano-forte', and all drinking 'the best wine of the Don', Clarke 'called to mind the erroneous notions we had once entertained of the inhabitants of this country; notions still propagated by the *Russians* concerning the *Cossack* people'. Who would have imagined finding portions of refined life in the hinterlands capable of entertaining English gentlemen! 'Perhaps few in *England*, casting their eyes upon a map of this remote corner of *Europe*, have pictured in their imagination a wealthy and enlightened society, enjoying not only the refinements, but even the luxuries, of the most civilized nations.'[114]

The Cossacks cultivated a life in their towns that few foreigners could have been prepared to expect. They were distinct people – not only in their origin (they were not, *pace* Hanway, a 'species of Tahtars'), but in their mode of life and customs, which Clarke's lengthy summary of the debates on their origin and the etymology of the name 'Cossack' was offered to prove.[115]

The ethnographic diversity was stupefying, rendered more complex with every traveller's description. By the late eighteenth century, grammarians and philologists had grown passionate to trace and record the diverse marginal and minority tongues in the European hinterlands. Explorers with an anthropological bent sought to bring to the present Gibbon's account of the ancient struggle between civilisation and barbarism among 'the wild people who dwelt or wandered in the plains of Russia, Lithuania, and Poland', and give detail to Johann Gottfried von Herder's account of Slavic folklore and analysis of the level of civilisation in the 'wild regions'.[116] Both were accounts produced in the 1780s: Gibbon's was erudite, but there were limits to what could be learned about the present from the annals of antiquity; Herder's was a typical philosophic history of the period. His history of humanity was painted with broad brush strokes, offering lots of colour but little detail. Clarke had the anthropological bent, and worked slowly through the territory, recording his observations with the patience and attention of an engraver. Observations on the inhabitants of the southern provinces provided more than ethnographic and natural historical information for the European atlas. They provided clues about the progress of Russian colonisation and the state of imperial rule and expansion. Accounts of the reconstruction of empires from Gibbon to Volney outlined the historical difficulties encountered in governing large empires. But how difficult was it to *civilise* as well as govern a growing empire? Furthermore, travellers provided observations for possibilities of extending East Indian trade routes. With every successive Russian conquest European attention to the prospects of accessibility to commercial seas – not only the Mediterranean but the Black and Caspian Seas – grew more acute. In the 1740s, Jonas Hanway had already written about the conditions of British trade through Russia over the Caspian Sea into Persia, and, in the 1770s, William Coxe addressed speculations about the 'revolution of trade in Europe' that would be caused by Russia's dominance in the southern provinces. They might control a profitable trade with the Tartars, Turks, and the Greeks in the Levant, but it depended on 'such casual circumstances as the coalition and rupture of rival and neighbouring powers'.[117] Monitoring the internal political economy and cavalry warfare in Eurasia was therefore in the commercial, as well as the philosophical, interests of Europeans.

Turks and trade

Between Russia's hustle to procure trading vents along the northern territories of the Black and Caspian Seas and the political tensions building throughout the Ottoman Empire were indigenous peoples working to secure wealth by creating their own Eurasian trade networks. Small, seemingly unassailable, communities punctured holes in the spheres of commercial interest spreading over the frontier territories. In particular, the Tartars (also 'Tatar') – members of several Turkish-speaking communities who dwelled along the Volga River and east to the Ural mountains – preserved a complex social organisation as well as civil and military independence from the Russians. The Tartars developed large classes of merchants, traders and a tradition of skilled crafts working with wood, ceramics, leather and metals. In the eighteenth century, their government head – the Kazan khanate – negotiated with the Russians a privileged position as commercial agents and administers of trading colonies in Central Asia.[118] Clarke, on his way to the port of Azov, encountered a number of merchant towns and commented on the varied picture of society and the buzzing Oriental bazaars which created a strong economic underpinning to the area.

Approaching an Armenian settlement, Clarke was struck by the mixed presence of 'Tahtars, Turks, Greeks, Cossacks, Russians, Italians, Calmucks, and Armenians'.[119] In the mid-eighteenth century, the Armenians had established their own thriving merchant culture along the Don which revealed their 'enterprising commercial spirit'. Instigated by 'commercial speculations', the Armenians travelled throughout different countries, from India to 'the most distant regions of the earth', to sell their commodities and manufactures. From their settlements along the Don, they supplied 'all the fairs of the neighbouring provinces; and these fairs afford the most extraordinary sights in *Europe*, because they are attended by persons from almost every nation. There is scarcely a nation, civilized or barbarous, which has not its representative at the fairs which are held along the *Sea of Azof*, and upon the *Don* . . .' Such commercial enterprise captured the spirit of all the territories of the Tartars. In the bustling fair at 'Nakhstshivan', just east of what is now Rostov-na-Donu, every trade in well-stocked shops was represented. 'Among other tradesmen, we observed tobacconists, pipe-makers, clothiers, linen-drapers, grocers, butchers, bakers, blacksmiths, silk-mercers, dealers in Indian shawls, &c.'[120]

These fairs thrived, but never appeared settled. With every fluctuation of the southern Russian frontier, each trading community shifted its position. It seemed to Clarke that the productive industry and commerce of the peninsula depended on their travels and migrant habits. True, the entire commercial culture was distinct from any European centre of commerce. It was common to invoke the classic association between settlement and civility as opposed to savagery and migration. Sailing along the Don, Clarke even suggested a European–Asian boundary. He contrasted Europe on the right with Asia on his left: 'the refinement, the science, the commerce, the power, and the influence of the one, with the sloth the superstition, and effeminacy, the barbarism, and the ignorance of the other.' Yet it was facile to brand tribes not connected to any particular territory with the imputation of barbarism. It ought to be confessed, thought Clarke, 'that the *peasant* of *Ireland*, the *smuggler* of *England*, or the *poissarde* of *France*, is altogether as unenlightened, more inhumane, and possesses more of savage ferocity, than either the *Laplander*, the *Tahtar*, or the *Calmuck*.'[121] The nomadic habits of the provincial tribes might even provide unique avenues towards civility, such as was demonstrated in the character of the Crimea Tartars. They annually embarked upon a 'pilgrimage to Mecca and Medina; so that a continual intercourse with other nations has contributed to their superior situation in the general scale of society'.[122]

Occasionally, economic motives and geopolitical pressures placed prejudicial views on others' place in the 'scale of society'. In the 1790s, with increasing competition between European powers and indigenous states, European trade merchants encouraged impressions – cloaked in free trade rhetoric – that the provincial tribes were barbarous and monopoly-mongers, slinging accusations against 'oriental despotism,' 'Muslim tyranny' and (somewhat awkwardly) 'the spirit of self-sufficiency'.[123] Clarke was familiar with the descriptive clichés, such as Montesquieu's opinion that 'the wild Arabs were a race of roaming thieves'. The roads leading to the Crimea 'are supposed to be infested with bands of desperate robbers,' wrote Clarke. Do not be fooled: 'Stories of this kind rarely amount to more than idle reports.' True, '*frontiers* are most liable to evils of this description', but from 'the author's own experience in almost every part of *Europe*, after all the tales he has heard of the danger of traversing this or that country, he can mention no place so full of peril as the environs of *London*.'[124]

Limits of enlightenment

Eighteenth-century encounters with the growing Russian empire and its European and Asian frontiers produced mixed accounts of their barbaric ancestry and its continuing struggle to civilise the fragmented, provincial, indigenous peoples. Two broad perspectives emerge in British travellers' accounts throughout the century. On the one hand, Russian society was ostracised for its Asiatic barbarity in contradistinction to Western, Occidental and European civility. On the other, some travellers admired its enlightened rulers who pushed through massive reforms, imitated the West and created a society well along the respectable path 'towards civilisation'. Many of the views were biased and prejudicial – from Clarke's criticism based on his uncomfortable metropolitan tour during Paul's paranoid reign, to Guthrie's devoted praise of his long residence in St Petersburg as court physician. Few views were judicial; Wilson's historical analysis was a uniquely balanced account.

An inherent tendency to be biased pervaded all accounts by authors appealing to Western models of civility to assess Russian culture. British critiques of the measure of Russian civility used categories invented by Enlightenment conjectural philosophy that gave Russia an automatic disadvantage. Civility was measured on progress in reproducing Western literature, classical scholarship and democratic reform. The 'yardstick' which measured civil progress was manufactured in the West. 'We are ourselves the supposed standards of politeness and civilization,' declared Ferguson. What then was to be learned from European encounters with people who possess features entirely different from the West's? In fact, much could be learned, and Ferguson went on to condemn conjectures that 'primitive savages' do not possess talent or virtue. 'Who would, from mere conjecture, suppose that the naked savage would be a coxcomb and a gamester? that he would be proud and vain, without the distinction of title and fortune?'[125] Yet these characteristics could be found in the indigenous societies of European frontiers.

The problems of Russian 'progress' were considered not only chronologically – determining an historical 'stage' of civility – but also demographically. Travellers did not determine a clear hierarchy of humanity, isolating serfs from slaves; savages from barbarians. There was no unified populace from which properly to assess 'a' Russian culture. In the West, the arts and sciences were well established and flourished in the metropolis. Intellectual foundations were rooted in physical settlement. 'The arts which pertain to settlement have been practised, and variously cultivated, by the inhabitants of Europe,' continued Ferguson. 'Those which are consistent with perpetual migration, have, from the earliest accounts of history, nearly the same with the Scythian or Tartar.'[126] Yet, as Clarke suggested, the comparison was not fair. The people in the southern and eastern provinces were, as their name suggested, 'wanderers'. They mounted resistance campaigns to avoid Western – western *Russian* – hegemony. Tartars resembled the past to Ferguson because they actively resisted Western notions of 'improvement'. As Peter I complained when he attempted to convince the Cossacks to abandon the sickle and use the scythe, 'though a thing be good and necessary, if it is new, our people will not do it unless forced to.'[127] Rather than dismiss the culture as 'savage', it seemed opportune to learn more about indigenous life and learn to trade on their terms. In this respect, any information gleaned from travellers that would help release Western traders from ignorance and dependency was welcome.

... Into Greece ...

While Antoine is cutting up an old goat to fry some chops in an earthen pan for Cripps' breakfast, I will make you a sketch of the luxuries we enjoy in Greece. It may cool your ardour for exploring these seas; for when I think of the enthusiasm with which I once planned such a voyage, it seems as a dream that vanished with the moments of repose. Danger, fatigue, disease, filth, treachery, thirst, hunger, storms, rocks, assassins, these are the realities! Will you believe, that even I have repented the undertaking? You once said all my letters begin with disasters. How can it be otherwise? I must shew things as they are. In my fourth decade, I can no longer scatter roses among thorns.

I call you to witness – was I always at sea a coward? Now the very sight of it sickens me to the heart. It has handled me so roughly that I shall never face it like a man again. Coming from Egypt, we tasted a tempest in a Turkish sixty-four; and since we were blown upon the rocks south of Naxos, to amuse ourselves with drying our rags, naked, upon a desert. But suppose it all goes well, and you have fine weather, and so on. Lice all over your body; lice in your head; fleas, bugs, cock-roaches, rats, disputing even to your teeth, for a crust of mouldy biscuit full of maggots. What's the matter now? 'Sir, we are becalmed!' Well, what of it? 'The pirates have lighted their signals, within two miles of us, if a breeze does not spring up, we are lost!' A breeze comes! it gathers force – it blows fresh – it whistles – it

111

roars – darkness all around – away goes the fore-sheet – the sea covers us – again a calm – again the pirates – Mercy! mercy!

Letter to the Rev. William Otter, two days before arriving in Athens. In W. Otter, *The Life and remains of Edward Daniel Clarke*, 2 vols (London: 1825), Vol. II, pp. 150–1

4

Southern Frontier: Greece and the Levant – The Archaeological Appropriation of the Historical Frontier

The rough seas and imminent danger that made Clarke's stomach curdle seems appropriate to the broader context of political turmoil in the southern Mediterranean that made for uneasy sailing. In 1793 the French National Convention (the new elective Assembly established by revolutionary leaders in 1792) declared war against Britain and General Napoleon Bonaparte's military status rose dramatically throughout the decade. But acute political manoeuvring by William Pitt's administration helped Britain to maintain maritime and commercial power, and an assertive position in regions being rapidly expropriated by Napoleon's army. In May 1798 Napoleon led approximately 38,000 French officers and soldiers, 16,000 sailors and some 180 assorted natural philosophers and artists on an 'Egyptian Expedition' with an aim to conquer the Orient, constrict Britain's trade routes and threaten its possession of India.[1] However, the British presence in the Mediterranean was not paltry, and Napoleon's entourage narrowly escaped British fleets before setting foot in Alexandria. News of Napoleon's invasion spread rapidly, and one alarmed British traveller residing in Constantinople, John Tweddell, wrote home wondering 'Where is Lord Nelson?'[2] In fact, as he wrote – in July 1798 – Admiral Nelson's flotilla loomed on the horizon and encamped in Cairo was General Hutchinson's squadron. Less than a month after Tweddell's query, Nelson attacked Napoleon's fleet in the battle of the Nile, forcing Napoleon's retreat and isolating his army in Egypt.[3]

In March 1801, British forces defeated the remains of Napoleon's army in Alexandria. In April, Clarke and his travelling companions sailed from Rhodes to Egypt, where they were able to rendezvous with Clarke's younger brother, George, who was a captain in the Royal Navy. With George, the travellers sailed into Aboukir Bay, surrounded by the smouldering remains of battle. Smoke pillared from the horizon, corpses floated in the sea and the stench of slaughtered soldiers, horses and camels permeated the hot air. 'Nothing could be more horrible,' wrote Clarke. 'The shores of Egypt might in truth have been described as washed with blood. The bones of thousands were whitening, exposed to the scorching sun, upon the sands of *Aboukir.*'4

The battles between the British and French that were fought on the southern shores of the Mediterranean marked an imperial frontier. Napoleon fashioned himself after Alexander the Great and was obsessed to conquer the Orient, to exceed the glory of the once great ruler of the ancient lands whom he so much admired. Beyond imposing military control, however, Napoleon also sought to establish an enduring legacy there by attempting to modernise the Egyptian government and promote the study of native culture. In Cairo, *savants* from the expedition established the Institut d'Egypte to help disseminate Western culture and ideas to the East, as well as explore and collect indigenous antiquities. Napoleon wanted his conquests to be not only tactical, but also beneficial to the entire act of empire-building. He wanted not only trade routes, but also the cultural artefacts that dripped the history of the ancient land's imperial heritage. As we begin to put into perspective the activities of the British travelling to Greece before and after Napoleon's defeat in Egypt, we will see that their own explorations and endeavours were likewise implicated in a type of archaeological imperialism.

During the revolutionary decades, both French and British commentators chose to represent ancient civilisation in such a way as to show that they were respectively the inheritors of the ancient principles of virtue, liberty and democracy. Like the conflicts on the battlefields, these commentaries were in competition with one another. For both, narratives about the ancient lands were invoked to make associations between the civility of the ancients and the self-defined civility of modern imperial rulers – the missionaries of the civilising process of the rest of the world.

Throughout the eighteenth century, both the French and the British developed strong cultural traditions in classicism and orientalism.

The British élite created for themselves a heritage where being educated meant learning ancient languages and taking the Grand Tour to Italy or maybe Greece: a liberal education at the 'ancient universities' of Oxford and Cambridge had as one pillar studies in the classics. Classicism also became a resource for justifying modern social and political structures. Ancient civilisation provided the principles upon which modern civilisation was founded and ruled by modern government. In the European contest over imperial domination in the late eighteenth century, the ancient lands were a profound focus: they were at once a territorial fighting ground and the *locus classicus* for defining the democratic and natural rights for different nations. Attention to the ancient lands, therefore, was manifest in a variety of ways, including historical accounts, travel narratives, antiquarian collections and paintings displayed in modern museums. One thinks of the immense influence of the controversial history of the *Decline and Fall of the Roman Empire* by Edward Gibbon or Constantine Volney's *Les Ruines*, which both invoked ruin imagery and the rational analysis of ancient civilisation as lessons for modern rule. What lessons were to be learned from the past in order to prevent the degeneration of modern civilisation? How should they be taught?

These sorts of questions were embodied in the long eighteenth-century quest to 'discover' the past. But it was only relatively late in the eighteenth century that Greece itself – rather than its history as revealed through ancient texts – became accessible to foreign travellers. Turkish rulers were protective of its territories, but after Nelson's defeat of Napoleon, the British easily established an alliance with the Ottoman Empire, a relationship that the British ambassador at Constantinople, Lord Elgin, would take full advantage of.

Models of antiquity

Since the seventeenth century, the 'ancients versus moderns' debate had raised the question of whether contemporary art and literature surpassed the achievements of the ancient Greco-Roman world. By the eighteenth century, Enlightenment reflections on the accomplishments of the modern epoch – in natural philosophy, printing and warfare – began to instil a sense of vast distance and difference between the ancients and moderns. Slowly, the commitment of Renaissance humanism to preserve the superiority and purity of a classical heritage (to which streams of neoclassicism and the later eighteenth-century 'Greek revival' can be connected) was increasingly doused by sceptics prodding the limits of ancient authority. A new sense of progress in history began to mark the modern distinct from the past; cumulative achievements in arts, science and literature provided more cultural capital – the touchstone of a society's worth – for modern civilisation. Historians and Enlightenment philosophers agreed that the present must learn from the past, but, for some commentators, the cultural accomplishments of the modern epoch could offer security and stability to modern states that ancients failed to achieve. Histories of antiquity and ancient civilisation, and especially accounts of the rise and fall of past empires, were encoded with lessons about how models of the past can help plan for the future.[5]

The concern presented to the British governing classes amidst these debates was how to prevent the growing British Empire from following in the footsteps of the Goths and turning into a ruined empire, allusions to which were all too familiar in travellers' narratives to the ancient lands. 'God grant the time be not near', prayed the educational author Thomas Sheridan, 'when men shall say, "This island was once inhabited by a religious, brave, and sincere people."'[6] His hope for the future was in maintaining the moral integrity and principles of modern government and cultivating the proper level of intellectual and material growth of the nation as well as the individual. History taught that the downfall of great empires was preceded by a laxity of public morals, and historians, reminiscent of Scottish social theorists, began to espouse theories of social progress which emphasised improvement of individual character as a means of preventing decay and ruin. Such was the history that began to feed into the ways that different European powers attempted to establish their new imperial identity and root themselves historically within a tradition of imperial reconstruction. Two eighteenth-century authors, one British and one French, whose works provided a literary backdrop to multiplying representations of the antiquity and the east were Edward Gibbon and Constantin-François Volney.

While sitting in Parliament and serving as a member of the Board of Trade in the 1770s, amidst debates about the framing of the Constitution and the administration of modern empire, Gibbon contemplated the historical development of jurisprudence, the arts, and education – the elements which combined to create a civilised culture. In a period when classical republicanism, which English theorists had been reconstructing since the Glorious Revolution, was being redefined by American revolutionaries, Gibbon reflected on his travels to Rome in 1764 where he had sat amongst the ruins of a city that was seen as the origin of the principles of civic liberty, humanity and civility.[7] His *History of the Decline and Fall of the Roman Empire* (1776–87) carried overtones of the Scottish Enlightenment's pessimistic implications that progress in civilisation could not occur without compromising individual freedoms and virtues, linked to the moral corruption and indolence of conquerors.[8] His work was a significant if also controversial contribution to the imagery of the new Whig history, which critically examined the development of ancient art, literature and topography as remnants of the imminent corruption and decay of empires. In the new historiography, comparisons drawn between classical and modern civilisation became virtually obligatory in political, literary, and artistic realms.[9]

In 1789, two years after the last volume of Gibbon's *History* was completed, the humanist cycle seemed to have rotated. The French Revolution appeared to be the culmination of a crisis in faith brought on amidst biblical challenges, evangelical millennial preaching and political criticisms. In 1791, the French philosopher and political historian, Constantin Volney, published *Les Ruines*, a meditation composed in 1784 while gazing on the ruins of provinces of Egypt and Syria, in which he also contemplated the inevitable decay of tyrannical powers.[10] Reminiscent of Gibbon's cautions and echoing the *philosophes'* theories of the degradation of human civilisation, Volney inquired:

> Who, said I to myself, can assure me that [Egypt or Syria's] present desolation will not one day be the lot of our own country? Who knows but that hereafter some traveller like myself will sit down upon the banks of the Seine, the Thames, or the Zuyder Sea, where now, in the tumult of enjoyment, the heart and the eyes are too slow to take in the multitude of sensations; who knows but he will sit down solitarily amid silent ruins, and weep a people inured, and their greatness changed into an empty name?[11]

Volney urged that the weakness of past empires lay in the government's corrupt principles, based on misguided notions of divine inspiration governing absolute rule, where tyrannical rulers attempted to hide their errors and vices by 'calling the delirium of their own misunderstandings the sacred mysteries of heaven'.[12] In a direct attack on the rulers of his homeland, he called instantly for a corrective of a new, free legislative system of universal rights, based on natural laws to govern man and society.[13]

Both works were extremely influential in late eighteenth-century debates over empire, invoking ruin imagery and the rational analysis of ancient civilisations as lessons for modern rule.[14] Especially after Pitt's India Act of 1784, encouraging judicial and political reforms in India, discussions over the 'fortunes' of empire extended beyond Westminster and into coffee-houses and parlours, not to mention lecture rooms at universities, where matters of imperial administration were pressed upon new generations of students. Lessons of how properly to manage the new empire and its vast territories were thus increasingly sought in histories of empire and ancient civilisation. Even attempts to stabilise the business of the East India Company appealed to models of expansion and decline of past empires, making knowledge of resources and economies of the ancient lands imperative to Company-bound administrators.[15] But the messages found in such histories varied according to the perspective and rhetorical strategy of the author. The political importance of Antiquity and the East towards the end of the eighteenth century reflected concerns ranging from securing British capitalist hegemony over eastern markets to the reassertion of national piety and the right to rule.[16]

But also by the later half of the eighteenth century, gradual exploration, archaeological discovery and growing familiarity with the Near East led to a radical revaluation of the uses of models of classical antiquity. From the 1750s, intellectual discussion increasingly generated particular discussion of ancient Greece. Exploration of Greece and the remnants of its past provided a variety of artists and writers the opportunity to address new ways that knowledge of ancient Greece could be used as a model for thinking about the present. Writing about the rise of arts and sciences, the Scottish political philosopher David Hume related his account of ancient Greece, 'a cluster of little principalities', with the evolving relationship of the neighbouring republics around Europe. His observation that 'EUROPE is at present a copy at large, of what GREECE was formerly a pattern in miniature', suggested a model of how cultural and commercial ties between neighbours have, and still might have been, developed.[17]

Lessons from classical Antiquity were also woven into the matrix of European civilisation through antiquities and artistic imitation. Artists found in classical scholarship continuity in the evolution of art and onward progress of civilisation. The founding of national museums in Britain and France gave institutional structure to shared discourses between history and art. The founding of the British Museum in 1753 was hot on the heels of the opening of an exhibition of paintings to the public at Luxembourg Palace in 1750. The democratic, encyclopaedic approach to the acquisition and classification of knowledge represented by these museums corresponded to the compilation of Diderot and d'Alembert's *Encyclopédie*, published from 1751, and the *Encyclopaedia Britannica* which began publication in 1760.[18] These institutions were faced with questions and criticisms regarding how to communicate lessons of history and morality to a wide public.

For Diderot and the French *philosophes*, history as well as collections and works of art carried moral messages. These messages were frequently drawn from classical themes. As the French artist Anne-Claude, comte de Caylus wrote, 'The monuments of antiquity lead to the expansion of knowledge. They explain unusual practices; they clarify obscure facts, . . . they put the progress of the Arts under our eyes, and they serve as models to those who cultivate them.'[19] Artists celebrated classical imagery in their own works. Charles-François Olier, Marquis de Nointel, a French ambassador who travelled the Levant, including Greece, in the 1670s, was one of the earliest French travellers to gaze in admiration at the buildings of the Acropolis and who also expressed a desire to collect its antiquities. 'The highest praise we can accord these original works,' he wrote, 'is to say that they would be worthy of a place in His Majesty's collections or galleries, where they would enjoy the protection extended by that great monarch to the arts and sciences which have produced them.'[20] Over a hundred years later during the revolutionary years in France, classical imagery provided artists with ways of asserting a new republican identity. Such were the series of paintings inspired by classical architecture and ruins that were prepared by the influential French landscape artist Hubert Robert for the re-planning of the Grande Galérie of the Louvre. Through Robert's works, *ancien régime* France was passing away into ruins, and a new social and political order was being erected. Or, in Britain, artists such as James Barry drew on classical themes to portray the progressive stages of human culture in paintings displayed by the

Society for the Encouragement of Arts, Manufactures, and Commerce in the 1780s.

But the work of the German art historian Johann Joachim Winckelmann was the most influential later eighteenth-century work to advocate that ancient Greece was the fount of classical civilisation and artistic originality. In his *History of Ancient Art among the Greeks* (1764), Winckelmann was the first to establish a comprehensive chronological account of all antique art, and proposed that Greek art should be the standard against which other art was to be judged.[21] In a sense, he invented the history of Greek art and devised a classification of Greek art based upon stylistic criteria. In his *History* and his earlier work, *Reflections on the Imitation of Greek Works in Painting and in Sculpture* (1755), he argued that Greek art was not influenced by a Roman tradition but that it was to be taken as original and canonical, an ideal to be studied and imitated. The aesthetic qualities of their representations of the human body were superior, he argued, because Greek culture fostered keener senses and attention to detail in the human form due to certain physical conditions in which they lived. The ancient Greeks were capable of producing exemplars of the human body in their statues because of their athletic culture and warm climate, which allowed them to shed their clothes and display their toned flesh.

Winckelmann championed the 'ideal of Greece' and imposed an aesthetic critique of ancient art upon his contemporary society that for decades was treated as the bible of neo-classicism. But he advocated an appreciation of ancient art through study and artistic imitation, not through pillaging the remains. He maintained that 'the only way for us to become great – indeed, if possible, to become inimitable – is to imitate the Ancients.'[22] Although, for all of Winckelmann's protestations against excavations in Greece, it was only through the rich collections that he encountered as Papal Antiquary to the pontifical court in Rome that he was able to write his accounts of Greek art. He never travelled to Greece, and in fact based his history of Greek art largely on the study of Roman copies of Greek originals.[23]

Winckelmann was a driving force behind the scholarly shift in interest from Roman antiquity to classical Greece, and his influence extended well beyond Germany. His theory of aesthetics was the centrepiece to the literary milieu that generated further intellectual pursuits of ancient Greece that flourished in Britain well into the early nineteenth century.[24] Contemporary political events added further charge to his theory of the development of ancient Greek culture. In the age of the American and French Revolutions, the pressures of confronting the demands for liberal democracy urged further explorations into ancient Greek democracy and government structure. But the political rhetoric of models of the past was increasingly based on new ways of growing familiar with the past. Winckelmann developed a theory of artistic history and ancient chronology that would be embraced by British travellers and collectors who had already begun their own empirical investigations into the ancient lands.

A Dilettante pursuit

From the fifteenth century, Greece had been under competing Ottoman and Venetian rule. With the eventual Turkish conquest in 1570–1 and its armies warring westward, travel to Greece by foreigners became exceedingly hazardous. Athens under Turkish occupation was strictly controlled; Christians were forbidden to set foot on the Acropolis without special permission. Since trade routes and pilgrimages to the Holy Land did not run through the Greek mainland, and with few adventurers willing to risk their lives for exploration there, the ancient land fell into relative obscurity.

Because Greece was largely inaccessible, Western attention turned to Rome, which had anyway conquered ancient Greece and shifted the centre of the classical world westward. In 1800, James Dallaway surveyed the major antiquarian collections in England and found that most – such as those of Earl of Leicester at Holkham, Robert Walpole at Houghton, the Earl of Egremont at Petworth, and Horace Walpole at Strawberry Hill – contained objects of Roman origin.[25] But some entrepreneurs did early on manage to tap other ancient resources. At the beginning of the seventeenth century, Thomas Howard, Earl of Arundel, travelled to Rome where he obtained papal permission to remove portrait sculptures, sarcophagi and altars from Roman ruins. Enthused by his growing collection, in 1621 he commissioned agents to search for more antiquities in Athens and Constantinople. The resulting cargo of Arundel's marbles amounted to over 200 Greek inscriptions – the first of their kind to reach England – and over 60 other statues, busts, and miscellaneous objects. His collection was deciphered and described in what has been called the first major work of classical scholarship in England, John Selden's *Marmore Arundelliana* (1628), which established the textual conventions for providing conjectural restorations of missing letters by printing them in red.[26]

A half-century later, a British gentleman who had been exiled from England during the Civil War and a French doctor-turned-antiquarian joined forces in 1675 and travelled together in the Levant. George Wheler and Jacob Spon spent near a year exploring ancient sites, including Athens in February 1676, where they made the first comprehensive survey of the ruins on the Acropolis. They were also the last to describe in detail the Parthenon before a Venetian shell exploded an ammunition reserve during an attack on the Turks in 1687, destroying most of the interior walls of the temple and columns on the north and south sides. After their tour, Spon was the first to publish an account of their travels, publishing *Voyage d'Italie, de Dalmatie, de Grèce et du Levant* in 1678, which soon after appeared in numerous translations and further editions.[27] Wheler's *Journey into Greece* was published in 1682. Spon was a scrupulous investigator who was concerned not just to uncover and examine new specimens, but to record in detail and fully document his findings, combining philology, epigraphy and archaeology in his analyses. Spon and Wheler's works were landmarks in Greek studies, and were standard sources throughout the eighteenth century. While Spon's small, dense volume was not as elegant as Wheler's lavish folio, he managed to pack more detail than Wheler and attracted more immediate attention to his account of their journey. But Wheler was by no means neglected. In 1820 Clarke was to publish a biographical account of him, praising his pursuits:

> Respecting the merits of Wheler, as a traveller, there can be but one opinion among those who have had the opportunity of judging. That he was diligent in his researches, intelligent, faithful, a good naturalist, and a zealous antiquary, cannot be disputed. That he was profoundly learned, will perhaps not be so readily admitted. It may be said, that for the erudition displayed in his book of travels, he was mainly indebted to his companion Spon; a charge easily urged, and after all not so easy to be proved. Wheler confesses, that he copied into his work some passages as he found them already published by his fellow traveller: but the facts, to which those passages relate, may have existed previously in his own journal.[28]

The qualities that Clarke here praised in Wheler – his diligent research, his faithful and ambitious antiquarianism and his abilities as a naturalist – marked the makings of a model traveller. Spon and Wheler have been described as the founders of modern Greek travel literature, but it took nearly a hundred years for English travellers to return to Greece with fresh enthusiasm.[29]

In 1734, a group of gentlemen connoisseurs recently returned from the Grand Tour to Italy transformed what had been an aristocratic London dining and drinking club into the 'Society of Dilettanti'. It was a Society dedicated to the cultivation of artistic taste, with special devotion to classical art and antiquities. For roughly the first twenty years of their existence, the Society fell short of fulfilling their cultural ambitions, although their support for the establishment of an academy of arts did help the eventual founding of the Royal Academy. By the 1750s, their patronage of the arts had expanded from Italian culture to the promotion of studying the remains of classical antiquity in Greece, to which the mid-eighteenth-century Hellenic revival owed much. As the late-Victorian historians of the Society put it:

> It was to the credit of the Dilettanti, that at the outset they recognised the true and guiding principle in classical archaeology, that the numberless monuments of sculpture, architecture, or painting which were continually being dug up in Rome, Naples, or the surrounding districts, were in the main but imperfect reflections of the pure light of Hellenic art and culture, the true source of which was to be found alone in the soul of Greece.[30]

In 1751, the Society agreed to support a project that would raise the promotion of Greek archaeological and architectural scholarship – of the Hellenic ideal – to new heights. Three years earlier the Society received a proposal from the painter James Stuart and the architect Nicholas Revett who argued that 'Athens, the mother of elegance and politeness . . . has been almost completely neglected, and unless exact drawings from them be speedily made, all her beauteous fabriks, her temples, her theatres, her palaces will drop into oblivion, and Posterity will have to reproach us.'[31] Their journey to Athens, where they were resident for nearly two years between 1751 and 1753, was subsidised by the Society.

While in Athens, they explored, excavated and drew indefatigably, establishing new standards of precision in representation. Their aim was to identify, measure and document the monuments and antiquities. Stuart drew and painted, while Revett measured and modelled the architectural ornaments. By the 1760s, the Society had raised thousands of pounds in order to support further expeditions to the eastern Mediterranean, and funded the publication of Stuart and Revett's detailed four-volume *Antiquities of Athens* (1762–1814; supplementary fifth volume in 1830).

In their Preface, they acknowledged that Italian antiquities had received due attention from skilled artists producing detailed engravings, but, by providing their own accurate representations, they humbly proposed that their work would prove acceptable 'to lovers of architecture, if we added to those collections, some examples drawn from the antiquities of Greece.'[32] From conception to finished product, this archaeological and architectural treatise, with engraved plates, maps and textual descriptions, took over eighty years to complete (neither author living to see the final product), but it became an enduring achievement as a respected work of reference, an unprecedented handbook to 'Grecian taste'.[33]

Other forms of contemporary literature also helped stimulate *Grecian gusto* of the period. The work sponsored by the Dilettanti was not long after the revival of Homeric studies, with Alexander Pope's translations of the *Iliad* (1715–20) and the *Odyssey* (1725–6), and not long before Robert Wood's critical examination of Homer in his *Essay on the Original Genius of Homer* (1769). But the Society provided new ways to generate classical scholarship through analysis of antiquities *in situ*. Among the scholars to benefit from the Society's travelling fund in the 1760s were Nicholas Revett (this time to Asia Minor), the historical painter William Pars and the Oxford don Richard Chandler. Their travels and the subsequent publications shed the sense of individual adventure that could characterise earlier accounts and were undertaken with new rigour in organisation and planning.

In the Preface to his *Travels in Asia Minor* (first edition 1775), Chandler began by reprinting his 'instructions' drawn up on behalf of the Society by Robert Wood. The aim of his 'journey into a remote country' (in fact two journeys which took place between 1764 and 1766), was to 'procure the exactest plans and measures possible of the buildings', to make 'accurate drawings of the bas-reliefs and ornaments' and to copy 'all the inscriptions you shall meet with'.[34] Chandler received £200 from the Society to defray all incurred expenses (altogether the Society provided £2,000 for the three travellers). His published travels prove he followed their instructions to 'keep a very minute journal of every day's occurrences and observations, representing things exactly in the light they strike you', unfortunately down to the last detail: 'without any regard to style or language, except of being intelligible'.[35] The travellers scoured all the major sites for classical ruins: Didyma, Miletus, Clazomenae, Erythrae, Teos, Priene, Tralles, Laodicea, Sardis, Philadelphia and Magnesia. His journals recorded their quests to find temples, and the arduous labour of climbing, digging and patiently transcribing. Their persistence paid off. Chandler handed his journals to the Society noting all the errors in previous published guides he believed he corrected, from Strabo's *Geography* (written in the first century AD but was still the standard reference for Greek geography) to Stuart and Revett, to Richard Pococke's *Description of the East* (London, 1743–5).[36]

But the ravages of weather and war often frustrated the search for the past. Looking for where the 'renowned Athenian' lived – where the *classical* Greek once lived – was made difficult by 'time, violence, and the plough'. On visiting Athens to study its past, Chandler advised, 'the traveller will regret, that desolation interferes, and by the uncertainty it has produced, deprives him of the like satisfaction; but, in the style of the ancients, to omit the research would merit the anger of the muses.'[37] Further, their efforts were not without risk to life or limb. While the Society's travellers used diplomatic links and bribes to procure a *firman* (a document from the Turkish authorities authorising their activities), they still faced robbery, disease and general discomfort. 'Our mode of living in this tour had been more rough, than can well be described,' was Chandler's frank assessment. He recalled that in July 1766:

> We had experienced, since our leaving Athens, frequent and alarming indisposition. We had suffered from fruits, not easily eaten with moderation; from fatigue; from the violent heat of the sun by day, and from damps and the torments inflicted by a variety of vermin at night; besides the badness of the air, which was now almost pestilential on this side of the Morea [of the six military divisions the Ottomans imposed to divide Greece]. My companions complained. Our servants were ill; and the captain, whose brown complexion was changed to sallow, had grown mutinous, and declared he would go away with his vessel, as he must perform a long quarantine at Zante, if his return were delayed; the annual unhealthiness of the Morea, toward the end of the harvest, requiring increase of caution, and the magistrates of the island restraining the intercourse with the continent at that season.[38]

But the revered remnants of the classical legacy – the products of the 'parents of philosophy' – warranted the toil of continued research.[39] For there the ruins sat, to the agitation of these scholars, neglected and half-buried in dirt! He, like previous scholars, lamented the condition at Sigeum of the famous Sigean Stone, believed to have powers against the ague: 'Above half a century has elapsed since it was first discovered, and it still remains, in the open air, a seat for the Greeks, destitute of a patron to rescue it from barbarism, and obtain its removal into a safer custody of some private museum, or, which is rather to be desired, some public repository.'[40] The Turks enforced an oppressive regime on the Greeks and, it appeared to these travellers, showed no interest in preserving the treasures of antiquity.

Under the aegis of the Society of Dilettanti the travellers asserted their privilege to be not only arbiters of taste, but saviours of monuments of antiquity. Prior to the mid-eighteenth century, the celebration of the Roman and Christian past had dominated historical scholarship, but by Chandler's time new attitudes and scholarship began to give priority to the 'Greek experience'. But it was a quest to discover *classical* antiquity, not modern Greece, which was thought to hold value for illuminating Western values. Those born and bred in the East were irrelevant – almost transparent – in travellers' accounts. The ruins were important; whether discussed in Britain, France or Germany, they instilled a reverence that far surpassed or supplanted concern over an Eastern present. As the French *philosophe* and encyclopaedist Denis Diderot exclaimed when he saw Hubert Robert's classical paintings in the Louvre, 'the ideas which the ruins awake in me are grand. Everything vanishes, everything dies, everything passes, only time endures.'[41] Conspicuously absent in such sparks of the philosophical imagination was any reference to the living ancestors of the ancient Greeks. For most, modern inhabitants simply cluttered the landscape. As the Society of Dilettanti explained in the Preface to Nicholas Revett and William Pars's *Ionian Antiquities* (1769), their researchers aimed to recover humanity's history through its antiquated remnants:

> The Society directed them to give a specimen of their labours out of what they had found most worthy of observation in Ionia; a country in many respects curious, and perhaps, after Attica, the most deserving the attentions of a classical traveller. . . . The knowledge of Nature was first taught in the Ionic school: and as

geometry, astronomy, and other branches of the mathematics, were cultivated here sooner than in other parts of Greece, it is not extraordinary that the first Greek navigators, who passed the Pillars of Hercules, and extended their commerce to the Ocean, should have been Ionians. Here history had its birth.[42]

It was most beneficially through topographical exploration that history was reconstructed. Activities on the mainland of Greece brought travellers still closer to the history of humanity. The Athenian 'parents of philosophy' were believed to have given birth to institutions, ethical values and ideas that were taken as points of reference and precedence for modern political rule. The offspring of the ancients were not the modern Greeks, however, but the children of the European Enlightenment.

Travellers considered the modern rulers of the ancient lands barbaric and ignorant of the glorious past that was trampled under their feet. The repressive Turkish regime put at risk attempts to acquire knowledge of the vestiges of civilisation; but at the same time, continued neglect of the ancient archives was considered by travellers an assault against Western heritage, since examining ancient Greece was in part a way for Georgian intellectuals to examine themselves.

Contempt for what they conceived of as native ignorance was repeatedly expressed. Chandler recalled that one day his Turkish guides requested he distinguish them by name – Mahmet, Selim and Mustapha – but their presence was never so particular, and in his narrative they remained 'the Turks' or 'the savages'.[43] Chandler's contemporary, Edward Gibbon, wrote of the blighting barbarism in Ionia that menaced the greatness of the familiar Roman heritage. 'The provinces of the East present the contrast of Roman magnificence with Turkish barbarism. The ruins of antiquity scattered over uncultivated fields, and ascribed by ignorance to the power of magic, scarcely afford a shelter to the oppressed peasant or wandering Arab.' Neither did the native Greeks themselves fare any better (see Plates 9 and 10). Again we read Gibbon, who described the modern Greeks as people who 'walk with supine indifference among the glorious ruins of antiquity; and such is the debasement of their character that they are incapable of admiring the genius of their predecessors.'[44] Even Winckelmann picked up on and repeated what travellers to Greece had said about their present condition:

> The modern Greeks, though composed of various mingled metals, still betray the chief mass. Barbarism has destroyed the very elements of science, and ignorance overclouds the whole country; education, courage, manners, are sunk beneath an iron sway, and even the shadow of liberty is lost. Time, in its course, dissipates the remains of antiquity: pillars of Apollo's temple at Delos, are now the ornaments of English gardens: the nature of the country itself is changed.... Unhappy country![45]

As late as 1820, publishing an account of his Greek travels on the eve of the Greek war of independence, the Cambridge don and Anglican clergyman Thomas Smart Hughes referred to the modern Greeks as 'that unfortunate race, occupants of the soil, if not the legitimate descendants of those heroes, whose names still shed a blaze of glory over the land which contains their ashes.'[46] Such judgements asserted that the modern Greeks bore no resemblance to their ancestors, and that their condition was only relevant in so far as their activities affected the preservation and research of the past.[47]

Was there hope for the future? Could the condition of Greek society be improved if their political oppressors were overthrown? At one point, Chandler briefly contemplated these questions. As if almost by accident, he looked up from the ground and realised that around him the modern Greeks were disquieted with their own political subjugation. 'They are conscious of their subjection to the Turk, and as supple as depressed, from the memory of the blows on the feet, and indignities which they have experienced or seen inflicted, and from the terror of the penalty annexed to resistance, which is the forfeiture of the hand uplifted.' But he held up little hope, for the 'body politic [was] weakened by division'.[48] His brief contemplation of contemporary political problems was almost lost in his directive to assess and describe the archaeological topography of Greece. But later commentators made such questions central to their concerns.

By the 1790s, British travellers brought new narratives of the condition of modern rule in Greece to bear on questions about the excavation and preservation of traces of ancient civilisation. Fewer were prepared to cast off carelessly criticisms of social degradation and intellectual ignorance. The *present* became directly implicated in concerns over the Hellenic ideal. *Time* was of the utmost importance, but not the ruins of time – the two millennia that had elapsed since Homer or Plato had pen in hand – but the pressures of time to penetrate, excavate and possess historical fragments before any other nation interfered. The Turks were not the only tyrants. The competition over the imperial frontier which travellers to the eastern Mediterranean encountered involved not only a military response, but a cultural campaign. The principles of enlightened government rule was not an abstract historical lesson, but a practical problem necessitating a response to Napoleon's attempted conquest of the East. As will be further discussed below, others also saw in these events the opportunity for a rally cry for a Greek revolution. Athens once humanised military conquerors, and it was thought by some that Britain had a political responsibility and social duty to promote the civilising process. But there was no unifying theory of how best to demonstrate the principles of a 'free government' as understood through associations with ancient models. As suggested earlier, not only the British, but in particular French and German travellers also developed interests in the ancient lands in attempt to seek out historical precedence for modern rule and support for claims that they were the true descendants of the Greek heritage. Beginning in the 1790s and continuing into the early nineteenth century, British travellers' accounts and activities in the ancient land must therefore be viewed in large part as reactions to imperial states at war.

Acquiring antiquities Part I: Imperial self-fashioning

In 1799, Thomas Bruce, seventh Earl of Elgin, was appointed British ambassador to the Ottoman Sublime Porte, a position he held until 1803. Since childhood, Elgin had cultivated a keen interest in classical art and learned that Greece possessed treasures superior to what was familiar in Italy. Encouraged by Thomas Harrison, the neoclassical architect who had worked on an estate of Elgin's in England, Elgin planned to expand his knowledge of Greek sculpture and architecture and make his embassy 'beneficial to the progress of the Fine Arts in Great Britain'.[49] To this end, Elgin began to assemble a team of painters and draftsmen with the intention of illustrating and producing plaster casts of Athenian architecture. Elgin had high ambitions and cared to spare no costs. Among the interested artists who were interviewed to join Elgin's crew were the watercolour painter Thomas Girtin, the aquatint artist William Daniell, the future architect of the British Museum Robert Smirk and his brother Richard, and, on the recommendation of renowned portrait painter Benjamin West, J.M.W. Turner.[50] Various problems in the negotiating process prevented any of these artists from participating, but on the advice of Sir William Hamilton, the British Ambassador at Naples and an enthusiastic antiquarian, Elgin hired Giovanni Battista Lusieri to head the artistic mission.

Lusieri met Elgin's secretary, William Richard Hamilton (later a trustee of the British Museum), and from Rome and Naples the two recruited additional artists, draftsmen and the appropriate materials for illustrating and making casts of statues and sculptures. Whether the original intentions were limited to drawing, copying and perhaps also collecting pieces that had already fallen is not clear. Before any actions could be taken, Elgin needed to procure a *firman* from the Disdar (Turkish military commandant of the Acropolis). This was eventually granted in early 1801. It submitted that, owing to Elgin's desire to 'read and investigate the books, sculptures, and others works of ancient Greek science and philosophy', permission was granted for his agents to examine, copy, model and measure the sculptures, ornaments and remains of buildings.[51] Such permission was not exclusive to Elgin, but, as we will see below, favourable relations between the Turks and the British eventually led to a more liberal interpretation of what rights Elgin's agents were granted. But, by the time they first arrived in Athens, in August 1800, a French team of artists and sculptors had already been at work on the Acropolis.

In 1780, a young French nobleman, Comte de Choiseul-Gouffier, and his travelling companion, Louis François-Sebastien Fauvel, toured Greece. Three years later, Choiseul-Gouffier was appointed French ambassador to Turkey, and he took that opportunity to appoint Fauvel as his agent in Athens, also obtaining permission for Fauvel to draw and make casts of the sculptures on the Acropolis buildings. However, Choiseul-Gouffier also had a hidden agenda and frankly urged his agent to collect as many original pieces as possible. 'Take away everything you can,' he commanded. 'Do not neglect any opportunity to remove everything in Athens and its neighbourhood that is removable.'[52] Fauvel managed to meet part of these demands, gathering a number of antiquities, and, despite being strictly forbidden to remove sculptures from the Parthenon, he also managed (by bribery) to obtain a segment of the frieze and a metope which were lying amongst the ruins.

The outbreak of the French Revolution slightly changed the course of events. Accused of conspiring to aid the royal family, Choiseul-Gouffier was declared a traitor to the Republic and went into exile in Russia. The new Republican government confiscated his possessions, including his collection of antiquities which was sent to France from Greece.[53] During these events, Fauvel continued to work in Athens, but now as employee of the French government, being instructed by the Commission of Monuments to care for the collection in Greece as part of national property. Fauvel's new instructions were part of a growing concern in the new French republic to conserve artistic and scientific objects and make them nationalised property, free for the public.[54]

It is largely in the light of the growing interest by the French revolutionary government in Greek artefacts (which affected Fauvel's activities in Athens in the 1790s) that the activities of later Britons (including Elgin) in Athens must be viewed. At the beginning of the nineteenth century, the battles in the ancient lands were conflicts not only over an imperial frontier, but competition over symbolic resources for historical legitimisation of modern 'democratic' rule. As one historian of French travel has noted, '[f]rom the time of the *fêtes gallantes* to the clash of arms on the battlefields of the French Revolution, the Greek world provided a wide repertoire of themes, images, and standards whose appeal varied from generation to generation, and from group to group. . . . Pressed for time and unable to replace the old order with a new one, the revolutionaries looked to the Greco-Roman republics as prototypes of liberty and patriotism.'[55] This is true not only of different ideological camps within France, but between British and French interests in Greece. Thus before proceeding to discuss the activities of British travellers to Greece at the turn of the eighteenth and beginning of the nineteenth century, it is necessary to outline the ways that the French republic promoted its own programme of imperial archaeology.

Classicism and the French republic

Interest in classical Antiquity took on a variety of new meanings soon after the new elective Assembly, the National Convention, declared a Republic (1792). The French Revolutionaries made new pronouncements about the democratisation of the arts and about the ways that museums and art galleries were thought to embody the principles of liberty and democracy. The uses of Antiquity in the arts were also transformed. Revolutionaries wanted to connect the new political order with a purer and nobler era of Europe's past; classical studies acted as a blueprint for returning to a golden age of liberty, equality and fraternity.[56]

Royal Academies of painting and sculpture in *ancien régime* France were essentially instruments of royal power, and the work of the students was commissioned through artistic patronage and court society, a controlled relationship which reinforced the class hierarchy, in which artists, like artisans, were subjects under monarchical rule. If not displayed in private estates, works of art found a place in a royal repository, off limits to the public. Soon after the establishment of a new political order and a new calendar, the Minister of the Interior, Jean-Marie Roland, wrote to the Republican *ideologue* and artist Jacques-Louis David, explaining the importance of establishing a new museum for the Republic: 'This museum must demonstrate the nation's great riches . . . the national museum will embrace knowledge in all its manifold beauty and will be the admiration of the universe. By embodying these grand ideas, worthy of a free people, . . . the museum . . . will become among the most powerful illustrations of the French Republic.'[57]

Under the new political order of the 1790s, art was to be produced freely and for the people. For many, the development of public museums would become central to the new political ideology. In 1793, the National Convention, led by David, abolished the 'Royal' from the Académies, and, although maintaining their previous classical educational principles, restructured them as an institute for public education (the Ecole Spéciale de Peinture, Sculpture, et Architecture). As Armand-Guy Kersaint imagined when writing about the triumph of the new regime over the old regime, Paris, 'peopled by a race of men regenerated by liberty', should succeed Rome as the 'capital of the arts'.[58]

Representatives of the Third Estate viewed themselves as the producers of wealth which the other estates had squandered, and the Revolution redistributed the property. The museum was to embody the idea of collective ownership, of shared wealth (released from the clutches of the upper estates), of free access, expression and freedom to display the fruits of their efforts. The citizen was to share a 'national character and the demeanour of a free man', asserted Abbé Henri Grégoire.[59] Soon after these declarations, the king was executed, the French revolutionary government declared war on England and Holland, and a 'Reign of Terror' was inaugurated; but at last, also in 1793, the Louvre offered free and open access to the public. As the Louvre's catalogue for that year proudly stated, 'The form of arts, like the political system, must change; art should return to its first principle – to the imitation of nature, that unique model for which unfaithful copies have so long been substituted.' The new Minister of the Interior and *philosophe*, Dominique Garat, announced that the goal of artists should be 'to instruct men, inspire in them the love of goodness and to encourage them to live honourably'. The arts should stimulate the 'moral regeneration' of the nation.[60]

French classical paintings and sculpture provided powerful imagery which was used to educate the people about the principles of the new political order. The famous Republican painting by Jean-Baptiste Regnault, *Liberty or Death* (1794–5), inspired by the revolutionary slogan, was hung in the debating chamber of the new National Convention. The iconography of the 'Republic', represented by a classically draped, working-class woman, would later be emulated by Eugène Delacroix. In his *Liberty Guiding the People* (1830), where she wears a red Phrygian cap to symbolise Liberty, and either carries a pike – the weapon of the people – or a bundle of rods carried by the attendants of Roman officials, which symbolised Fraternity. The Republic overthrew monarchical rule, where citizens had laboured under a regime, and where artistic innovations were not free but controlled and patronised for particular, patrician ends.

However, during the Terror, political and military factions within France fragmented Republican ideology. Dominique Garat had stated in July 1793 that the museum was intended to show 'to both the enemies as well as the friends of our young Republic that the liberty we seek, founded on philosophic principles and the belief in progress, is not that of savages and barbarians.'[61] The message was that the quest for civility, freedom and equality should not be forgotten during times of revolution and war. The development of museum collections was intended to demonstrate this rationality. But, the following year, in June 1794, the official and systematic confiscation of art by French troops was authorised by the Committee on Public Instruction. This enabled General Napoleon Bonaparte, particularly after his later Italian exploits, to add considerably to the collections of the Louvre. However, while reaping these rewards, by adorning Paris with exotic cultural gifts, Napoleon's activities and collections also worked to refashion France's imperial image. He used the arts to express his own political ideology of conquest and rule, foreboding images for his 18th Brumaire coup in 1799 (year VIII).

Napoleon's imperial identity and artistic propaganda

Napoleon's ambitious military missions acquired collections which were used to fashion a new imperial identity for France. War booty and prizes from his campaigns were used as ornaments to his new imperial regime; classical portraits now provided images of historical precedence for Napoleon's dictatorship. While artistic institutions were freer than under the academic stranglehold of the *ancien régime*, Napoleon exercised a different kind of control over them, using 'national' collections to express the ideology of the ruler. Napoleon's military campaigns were highly successful, and he made much of the 'trophies of conquest' that were won from these exploits.

Napoleon's Italian campaign of 1796–7 had early on reaped notorious rewards. In 1798 a 'triumphal entry' festival celebrated the return of soldiers who proudly waved their tricolour flags and hauled a wealth of new acquisitions to the Louvre.[62] The public then found themselves in a skilfully organised gala in admiration of captured treasures from antiquity. The carnivalesque procession encouraged spectators to cheer, not for their new rights and liberty, but to promote reverence for Napoleon's military prowess. One contemporary observer commented: 'The national museum and its precious contents are recompense for the lives and blood of our fellow citizens spilled on the field of honour. French artists are worthy of this prize; they fully recognise its importance.'[63] This description draws attention to a major shift in representations of civility: from freedom to domination, from creation to appropriation.

Napoleon had visions of grandeur. His ambitions were considered by some to undermine the spirit of moral regeneration and proper democratic rule. When Jacque-Louis David refused to join Napoleon on the Egyptian expedition in 1798, he did so with the lament: 'O well, I always did think that we weren't virtuous enough to be republicans.'[64] The next year, he displayed his *Intervention of the Sabine Women*, a portrait which drew on the theme of reconciliation, apropos of post-'reign of terror' sentiments. This painting is significant not only because of the association it made between the origins of ancient Rome and modern Paris, but because here David was clearly rethinking the principles of morality and virtue.

Napoleon's Egyptian expedition and the antiquities he had sent to the Louvre represent that a change in Republican ideology had occurred. The mentality expressed from the beginning of the revolution in 1789, which led to the founding of a national museum as a temple to liberty and a symbol of redistribution of property, turned, with increasing military might and successive campaigns, into a repository for 'trophies of conquest'. The emphasis on accessibility, instruction and cultivation of artistic freedom was replaced by an obsession for collecting tokens of memorable exploits as tributes to the French who had shed blood for the honour of their country.

When Napoleon attempted to conquer the East he likened himself to Alexander the Great, whose land and empire it had once been. But when he was defeated in Egypt and his expedition quashed, the appropriation of classical imagery to support a political discourse of freedom and republicanism was also to be turned on its head. If Paris was fashioned a new Rome, then London was to be the modern Athens. By the time Elgin's agents arrived in Athens, Fauvel, along with other French nationals, were arrested and imprisoned. Seizing all the apparatus geared for moving large marbles that Fauvel left behind, Lusieri and a number of other British travellers were set to pursue their own antiquarian interests.

Acquiring antiquities Part II: The 'Levant lunatics'

After Napoleon's defeat, British travel writers and diplomats began to appropriate the language used in the French republic regarding artistic and personal freedoms to legitimise their own pursuits in the East, and in attempt to demoralise Napoleon's troops. In establishing their new imperial identity, the British also searched for ways to root themselves historically in a tradition of imperial reconstruction. Just as the French Republic had fashioned an imperial identity by looking back to a Roman ancestry of democracy and liberty, the British developed their own particular interests and scholarship regarding ancient civilisation in the far shores of the Mediterranean.

After the battles ended and territorial redistributions started, a conflict regarding the ownership of war booty arose. Specifically, it was a dispute that concerned the 'law of prize', the sixteenth article of the Capitulation of Alexandria, which decreed that all natural history and antiquarian collections possessed by the French must be handed over to the British. When Clarke and Cripps arrived in Alexandria just after the Capitulation, Clarke offered his assistance to General Hutchinson to help procure any antiquity that could potentially be relocated either to Cambridge or the British Museum, 'as I know full well,' Clarke wrote to a friend back home, 'we have better Orientalists than the French.' Hutchinson sent Clarke, who at that time was travelling with Lord Elgin's secretary William R. Hamilton, along with members of the Society of Antiquaries (of London), to negotiate with the French General, Jacques-François de Menou. While 'backed with a menace from the British General, that he would break the capitulation, if the proposals were not acceded to,' Clarke and the others were instructed to 'discover what national property . . . was in the hands of the French.'[65]

Clarke's judgement that the British 'had better Orientalists' than the French was a way of invoking intellectual, rather than military, adeptness to justify their right to seize war booty. The claim that the British *knew better* than the French how to appreciate the antiquities – who were able to read, interpret and protect them – also suggested that the British possessed more highly cultivated skills for understanding classical antiquity. The competition for possessing relics of ancient civilisation was not merely about possessing property, but having the historical *right* to do so. That right was asserted in the argument that whoever had the ability to understand and more accurately reconstruct a portrait of the past, should inherit the heirlooms of their ancestors. It was, in essence, by virtue of their apparent ability to *know better* how to analyse Antiquity that enabled them to claim possession of the past. This sentiment also helps to clarify whose nation was being referred to when the British sought to find out what 'national property' was in the hands of the French. It might seem to refer to the victor's nation, who by wartime rights could claim conquered territory and all in it. But because many of the objects in question were relics of an ancient civilisation and an empire that had crumbled, possessions from an *ancient* nation symbolically represented the new property, revived power and hereditary continuity of a modern imperial state.

What Clarke and the others found when they were sent to sort out the 'Great Dispute' between Hutchinson and Menou immediately aroused much interest. It appeared that a French soldier, while constructing encampments in a mountainside, had chiselled out a 'triangular rock', referred to as the 'Hieroglyphic Table', or, today, as the Rosetta Stone. After some negotiation, Clarke, Hamilton and the representatives from the Society of Antiquaries secured the stone, which was quickly put on a ship and sent to the Society in London. But that was only the beginning of the appropriation of antiquities from hands of the French and from the ancient lands. The pinnacle of classical studies and antiquarian collecting soon became Greece. For Clarke, arriving in Greece in late 1801 marked the culmination of his travels through the frontiers of Europe. 'It is necessary to forget all that has preceded – all the travels of my life – all I ever imagined – all I ever saw!' he enthusiastically wrote to his colleague in Cambridge. 'Asia, Egypt – the Isles – Italy – the Alps – whatever you will! Greece surpasses all! Stupendous in its ruins! . . . Nothing ever equalled it – no pen can describe it – no pencil can portray it!'[66] Perhaps it was the predilection that the antiquities of Greece could not be 'portrayed' that large-scale collecting seemed to become the order of the day (see Plate 11).

When Clarke and Cripps arrived in Athens, Lusieri showed them around the Acropolis. By that time, Elgin's agents had managed to procure an extension to the original *firman* they were granted. Now they were no longer limited to painting and taking casts of fallen statues, but were permitted to erect scaffolding around the ancient temples, and the local authorities were forbidden from interfering with any agent who might 'wish to take away any pieces of stone with old inscriptions, or sculptures therein'.[67] Workmen were busy hanging ropes and pulleys, preparing to remove parts of the frieze from the Parthenon. Clarke confessed to the readers of his *Travels* how, when watching the Parthenon being dismantled, he and his party 'would gladly see an order enforced to preserve rather than destroy such a glorious edifice'. And to re-evoke the emotional distress and sorrow shared by the visitors to the Acropolis, he went on to recite a story about how, in the process of removing a metope, they accidentally dislodged adjacent masonry. Down 'came the fine masses of *Pentelican* marble, scattering their white fragments with thundering noise among the ruins.' Following this disaster, the local Disdar, already stripped of his authority to interfere by order of the *firman*, could only shed a tear and utter 'the end!' (see Plate 12).[68]

This provocative tale has often been cited to show that Clarke, a contemporary observer and influential British travel writer, opposed Elgin's activities on the Acropolis and the removal of the marbles to Britain.[69] It is true that Clarke added a long footnote to that part of his narrative where he expressed concern over the destruction of the Parthenon, and wondered why the same effort to remove the marbles might not have been deployed in urging the Turkish authorities to preserve them. But it should be remembered that Clarke presented his account of Athens to the public in 1812, when philhellenism was in the ascendant, another generation of British travellers had forged new relations with the Greeks, and the exploration of Greek topography had grown to be more comprehensive and sophisticated. By the 1810s, it was in keeping with current intellectual debate to offer a perspective of the relative merits of removing or preserving Antiquity from the ancient lands.

Clarke's apparent disapproval was prompted by a pamphlet penned by Elgin's secretary, William R. Hamilton, titled 'Memorandum on the Subject of the Earl of Elgin's Pursuits in Greece'.[70] There, Hamilton argued that the principal justification for the removal of the marbles was because of the sustained injuries to the statues due to the 'zeal of the early Christians' and Turkish artillery explosions. Clarke disagreed, and argued that the dilapidated appearance of the buildings and statues was due to the 'decomposition of the stone itself, in consequence of the action of the atmosphere over so long a time'.[71]

But Clarke did not totally reject the act of collecting antiquities abroad. After all, during his travels he shipped home something in the order of 76 crates of collectables. In fact, some of the largest came from Greece, including the Eleusinian statue thought to be Ceres, the Goddess of Agriculture, which, against protest of local farmers, he bribed (with the gift of an English telescope) the local authorities for permission to remove.[72] Removing the statue was essentially depriving the Greeks of the goddess who they believed elevated them above barbarians. As Richard Chandler had earlier explained, the Greeks believed that the goddess conferred two unparalleled benefits on ancient Greek society: 'the knowledge of agriculture, by which the human race is raised above the brute creation, and the Mysteries [secret rites], from which the partakers derive sweeter hopes than other men enjoy, both as to the present life and to eternity.' The statue represented a memorial to the bounty of the region and knowledge of the resourcefulness of the land. 'It has been asserted that the [Eleusinian] mysteries were designed to be a vehicle of sublime knowledge, and represented in a kind of drama of the history of Ceres "the rise and establishment of civil society, the doctrine of a future state of rewards and punishments, the error of Polytheism, and the principle of unity...".'[73] What Clarke captured for the University of Cambridge was a symbol of progress for civil society, more appropriate (he thought) for his *alma mater*, where piety and agricultural wealth provided the backbone to a liberal and enlightened education.

Clarke justified the removal of antiquities for their preservation based partly on his mineralogical theory of the decomposition of stone, and partly for his commitment to a pedagogical ideal of being able to display at home examples of art collected from abroad, at no cost to the public.[74] Clarke and Elgin were by no means the sole participants in the *Grecian gusto* that swept through the end of the eighteenth century. Others travellers from Cambridge University who visited Greece in the 1790s had similar interests in collecting classical remains. John Morritt, a graduate of St John's College in 1794 (later 'Arch-Master' of the Dilettanti Society), travelled through Greece and Asia Minor in 1794–6 with the Oxford scholar James Dallaway. In Athens, they met Fauvel, who, as well as 'protecting' the collections of the French government, also started his own trade in classical antiquities and supplied the British travellers with their own small collection.[75] John Tweddell was another Cambridge graduate who was to meet Fauvel in 1799 and assemble what he considered 'the most valuable collection of drawings of this country that was ever carried out of it'.[76] Alas, Tweddell was taken ill and died that year in Athens.

Already in Athens when Clarke arrived in 1801 were the Cambridge scholars William Wilkins, Edward Dodwell and William Gell (later Sir William). Wilkins had received the Worts Travelling Bachelorship, a yearly award for Cambridge graduates which enabled them to engage in scholarly travel for one year. An architect by training, Wilkins was predominately interested in bringing home designs and plans of Greek temples and was full of praise for Elgin's work in removing the friezes of the Parthenon to Britain.[77] Dodwell's trip to Greece in 1801 was his first; he returned in 1805 and 1806. His account of his tours to Greece was dense, and much indebted to previous scholars, his favourites being Chandler and Clarke.

Dodwell's narrative is full of inescapable reverence for the past and a land that was filled with the presence of the immortalised ancient authors.

> Almost every rock, every promontory, every river, is haunted by the shadows of the mighty dead. Every portion of the soil appears to teem with historical recollections; or it borrows some potent but invisible charm from the inspirations of poetry, the efforts of genius, or the energies of liberty and patriotism.

The classic shores of Greece overwhelmed his classical education; he walked though the land of all the poets, historians and orators 'to whom Europe has been indebted for so much of her high sentiment', and saw the remnants of Antiquity that even to his day served as 'models of perfection, and the standards of taste'.[78] But walking through the remnants of the land of lost gods was more than paying homage to ancient muses. These travellers set themselves the more exacting historical task of finding in a crumbling landscape a reliable topography of ancient civilisation which could be compared to stacks of written accounts.

William Gell, who graduated from Cambridge in 1798 (knighted in 1803), became a distinguished classical scholar after his travels through Greece and the Ionian islands on various trips between 1801 and 1806. Gell's publications range from topography, to ancient costume, to narratives about ancient custom and current manners.[79] Through his detailed topographical and geographical accounts of the ancient lands he gained the reputation as 'Classic Gell' (so nominated by Byron in his *English Bards*). That, however, was soon surpassed by the assiduous researches of the British army officer William Martin Leake in the first decade of the nineteenth century. His seven dense volumes, *Travels in Northern Greece and the Morea*, in addition to his volumes on the topography of Athens and Attica, earned him the epithet 'the founder of the scientific geography of Greece'.[80]

Most of these turn-of-the-century travellers from Cambridge University lost the romance of desolation of the mid-eighteenth century and prepared for their travels believing that a liberal education was becoming manifest in classical scholarship. It has recently been argued that mathematical studies in post-Newtonian Cambridge came to eclipse classical studies during the eighteenth century, a perspective that finds some support from the priority given to mathematical training in conferring degrees in the eighteenth century (classical studies were not integrated into the examination system until the 1820s). But in the making of the literary traveller, we begin to see an alternative approach to a liberal education. It has been pointed out, for example, that the number of books printed under the auspices of the University between 1791 and 1800 in the fields of theology and classics, compared to mathematics and natural philosophy, held a ratio of three to one.[81] In this milieu, which generated classical scholarship in a whole range of museological studies, travellers' perceptions of ancient lands were particularly astute. As the editors of *Museum Criticum; or, Cambridge Classical Researches* advertised, they welcomed any contributions to their publication that were 'original communications, such . . . as may contribute to throw any new light on the manners and customs, arts and sciences, history and antiquities of the Greek and Roman empires, whether from the observations of modern travellers, or the stores of ancient learning.' In short, the object of their 'literary criticism' was to pay 'critical attention to the languages of antiquity'.[82] The group discussed above (in addition to other travellers, such as Thomas Hope, Charles Kelsall and John Eustace) have more particularly been identified as 'Cambridge Hellenists' or 'Cambridge Graecophils', marking them as the leaders of a broader philhellenistic movement.[83]

Yet while these travellers were all committed to the promotion of classical scholarship through travel, mapping and collecting, not all agreed in how to determine the 'value' of the controversial 'Elgin marbles'. Some believed that the best way to convince Parliament to spend thousands of pounds on purchasing the collection 'for the nation' was to prove that England's possession of them would protect the marbles from the tyranny of foreign despots. Through these debates we can get a sense of how critics perceived ways that classical studies fed into imperial narratives.

To patronise or pilfer?

In the first two decades of the nineteenth century, British travellers continued to flock to Athens. Henry Holland, the Edinburgh physician who arrived in Greece for a two-year stay in 1811, noted that the proportion of British in Athens was ten times that of the French and Germans, and they took Athens 'as the centre of resting place to more extensive research'. 'The English', he observed, 'more than any other people, have cultivated the ancient, through the modern Athens.'[84] A contemporary traveller, Lord Byron, was not overjoyed to see swarms of his compatriots, whom he dubbed the 'Levant lunatics'. 'Athens is at present infested with English people,' he scoffed.[85] Yet the new century brought on new opportunity and desire for Greek travel. On the one hand travelling through the land offered a licence to literary acclaim, on the other it was a land to be surveyed and protected by the British. In British imperialist discourse at the turn of the eighteenth century, no one else qualified as the modern caretaker of civilisation and protectorate of its ancient heritage. The British presence in Greece was rationalised as an act of promoting freedom and independence for its indigenous population.

The rhetoric of freedom and independence, of overthrowing

subjugation and tyranny, was deployed as an antidote to the misguided principles that apparently governed the eruption of post-Revolution militarism in France. For British diplomats with an interest in Greece, Napoleon was held responsible for corrupting the values that promoted the search for the origins of democratic ideals in Western society. Edward Dodwell, visiting Ithaca after being released from French captivity following the Battle of the Nile, found irony in the zest to express enthusiasm for national freedom and democracy. He found traces of the republican motto 'Vive la république; liberté, egalité, et fraternité' scratched into the sides of a fountain in Ithaca when the French occupied the island in 1798. One irony was that these were principles that appeared to have been betrayed in Napoleon's feverish campaign of conquest. Liberty, equality and brotherhood appeared (at least in British perceptions of their imperial rivals) to have been replaced by sentiments of tyranny and domination. The fight for freedom from the shackles of monarchical oppression seemed to have changed into fights for extended rule. The other irony for Dodwell was that the modern Greek who quenched his thirst from the fountain was 'little conscious of being surrounded by such sublime conceptions'. The French seemed to have betrayed their own principles of attaining liberty and equality; the Greek, suffering 'the dark cloud of oppression which hangs over them, and which cannot be entirely overcome by the accumulated tyranny under which they groan', seemed to have forgotten the meaning of such political principles.

The political debates about national liberties and freedom extended beyond the battlefields and into the domain of artistic discourse. The urge to remove Greek antiquities was at one level driven by the desire to recapture and 'liberate' the past from the tyranny of the French. Such a rationale began to imitate the pronouncements for the overthrow of Turkish tyranny and the assertion of Greek independence. Some arch-pundits of the antique even linked the benefits of supporting Greek independence and overthrowing Turkish rule for acquiring more antiquarian collections. Such eager hopes were dismissed by William Gell:

> Those who vainly flatter themselves that the destruction of the Turkish barbarism would open to them the road to the investigation of Grecian antiquities, treasures of sculpture, and a new area of the arts and humanity, may assure themselves that no such effect would be produced. A long reign of anarchy would

be succeeded by a fresh and more active tyranny, during which, of Greek rules, strangers would be excluded, and the monuments of antiquity, fetching no price, would find their way to the lime-kiln.[86]

One underlying assumption of Gell's was that there in fact existed little chance of Greek emancipation, an issue further discussed by Henry Holland. Holland heard from a celebrated Greek physician named Ioannes Velara that his countrymen were tired of the neglect they had received 'from the civilized communities of Europe', and characterised Greek political sentiment in three ways:

> The insular and commercial Greeks, and those of the Morea, attached themselves to the idea of liberation through England; a second party, in which he included many of their literary men and continental merchants, looked to the then existing power of France, as a more probable means of deliverance; while the lower classes, and those most attached to their national religion, were anxious to receive the Russians as their liberators.[87]

From here their discussion went on to the 'comparative merits of the ancient Greeks and the civilized nations of modern Europe', during which Holland grew impressed with Velara's 'accurate understanding of the ancient authors' and his appreciation of the 'former glories of his country'.[88] But with the new favoured position in Athenian society that the English enjoyed in the early decades of the nineteenth century, more British began to believe that their presence encouraged steps to the liberation of both ancient and modern Greece. The two, however, were not taken together, and Holland for one remained sceptical about whether modern Greece could be transformed to match the civility of the rest of Europe. 'It still remains a matter of interesting speculation', he concluded, 'whether a nation may not be created in this part of Europe, either through its own or foreign efforts, which may be capable of bearing a part in all the affairs and events of a civilised world.'[89]

Others who held similar views argued that it was the duty of the British government to remove relics of antiquity from the neglect of barbaric Turkish rulers and misappropriation by French tyrants. As in the case of the purchase of the 'Elgin marbles', such reasoning was used by a variety of diplomats and artists who attempted to assign 'value' – both moral and pecuniary – to the collection.

In 1807, Lord Elgin – having been detained for three years as a prisoner of war in France when the brief peace between England and France abruptly ended in 1803 – finally returned to London. At his Mayfair mansion he had a 'shed' built in which he put his marbles on semi-public display. Attention to this massive collection was immediately aroused. Praise was poured forth by visitors, who included Benjamin West ('sublime'), Benjamin Robert Hayden ('My heart beat!') and Henry Fuseli ('De Greeks were godes! de Greeks were godes!').[90] Artists appeared to be largely favourable to Elgin, but rival collectors, such as the members of the Society of Dilettanti, underplayed their value. In 1815, the marbles were the subject of discussion by a Parliamentary Select Committee regarding their potential purchase 'for the nation'. After listening to the testimony of a series of travellers, artists, architects and dealers on the value of the marbles, a price of £35,000 was offered to Elgin for transferring ownership and placing them in the British Museum.[91] Elgin begrudgingly accepted this offer, complaining that his expenses in removing the statues came to double that amount. Besides that, he believed that the marbles demanded a higher estimation of value given their superiority to any French collection. He even paid Ennio Visconti, an Italian antiquary who in 1814 was working as a museum curator for Napoleon, to offer his own assessment. Elgin hoped that Visconti, whom he called 'the best judge in Europe', would convince the British government 'that the collection is highly desirable, and consider'd so by such authorities, as are conversant with Bonaparte's Collection . . .'[92]

The British valuers, however, had their own criteria to determine the value of the collection. For them, the 'value' depended not on the costs and problems of transport from Greece, Egypt or Constantinople to England, but on the potential *benefit* that the marbles could have for aspiring British neoclassical artists: the effect that these marbles would have on judgement and taste. The debates surrounding the publicity of the Elgin marbles in England had to do with issues ranging from legitimising the seemingly imperial act of 'raping' the 'land of lost Gods and men' of their material possessions to how aesthetically pleasing they were relative to other art forms. Although other commentators remarked rather sarcastically that the sheer benefit of the Elgin marbles had to be their public accessibility, 'so that the traveller who has in vain looked for them in Greece might at last find them in England!'[93]

The artists who testified to the Select Committee argued that

London should house the sculptures to demonstrate Britain's commitment to preserve the integrity of the pursuit of the arts.[94] The reason why Greece produced such extraordinary art in the ancient world, it was argued in 1816, was because it was promoted within a free government. As the Report of the Select Committee declared:

> Your committee cannot dismiss . . . how highly the cultivation of the Fine Arts has contributed to the reputation, character, and dignity of every government by which they have been encouraged, and how intimately they are connected with the advancement of every thing valuable in science, literature, and philosophy. . . . If it be true, as we learn from history and experience, that free governments afford a soil most suitable to the production of native talent, . . . no country can be better adapted than our own to afford an honourable asylum to these monuments of the school of Phidias, and of the administration of Pericles.[95]

In London, as in ancient Greece, the arts should flourish under a free government; they should not be subjected to the tyranny of French (or their associated predecessors, Roman) rule. Of course, 'free government' was a concept that was also meant to include liberal patronage for the artistic community. Money spent on marbles was also money for those to look after them and promote their presence.

The idea that the British government would demonstrate principles of liberty by buying Elgin's marbles and therefore supporting the pursuit of the arts was a notion imported from the eighteenth-century German art critic, Johann Winckelmann. In Germany, Hellenists also fostered similar anti-French sentiments. As Winckelmann commented, 'the French are irredeemable: classical antiquity and the French are at opposite poles.'[96] In his *History of Ancient Art Among the Greeks*, Winckelmann proposed that '[t]he independence of Greece is to be regarded as the most prominent of the causes, originating in its constitution and government, of its superiority in art.'[97] Having the Greek marbles in London, the new symbolic centre for freedom and the promotion of art, would enable British artists to have an established guide with which to evaluate their own art.

Indeed, it was this line of reasoning that another British traveller to Greece, John Cam Hobhouse, used to argue in favour of the removal of Greek antiquities to Britain. In the first volume of his published travel narrative, he emphasised in a long footnote his belief that the removal of the Greek marbles would benefit 'an infinitely greater number of [British] architects and sculptors' if they were in Britain rather than Greece and certainly France.[98] Not every British artist could make the Grand Tour, and having the sculptures in Britain would accommodate their interests. But others strongly disagreed, including Hobhouse's travelling companion, Lord Byron, who declared: 'I oppose, and will ever oppose, the robbery of ruins from Athens, to instruct the English in sculpture (who are as capable of sculpture as the Egyptians are of skating).'[99] Looking briefly at Byron's response to the appropriation of Greek antiquities reveals ways that the language of 'free government', rescuing the past from imperial tyranny and conquest by the French, could also be exposed as nothing short of Britain's own desire to symbolically dominate the past.

When Byron returned from his pilgrimage to Greece in 1811, he did not intend to suppress his opinions about the activities of the other British who had been resident in Athens for the past decade. On his journey home from Greece on board the *Hydra*, Byron, along with Hobhouse, sat among crates packed by Lord Elgin's agents which contained the last shipment of the marbles. Not long after disembarking in England, Byron penned a harsh letter to Elgin: 'I knew all about his robberies, & at last have written to say that . . . it is my intention to publish (in Childe Harold) on that topic . . .'[100] *Childe Harold's Pilgrimage*, the first two cantos of which were published in 1812, was Byron's metaphoric account of his travels in Greece. The poem poignantly conveyed his attitude concerning the collecting habits of the British élite and the national interest in the neoclassical movement.

His poem represents one of the first public displays of his life-long interest in Greek independence, and presents a glimpse of his anti-imperialist demeanour and radical politics. However, in the context of early nineteenth-century literary production, Byron's was only one of many contributions to the discourse about the 'anxieties and insecurities' of imperial policy.[101] Having been a student at Trinity College, Cambridge, Byron could be nothing but acutely aware of the cultural significance that representing the far shores of the Mediterranean could have. Byron's criticisms are mentioned here to suggest not only the diversity of media through which themes of classicism and imperialism were expressed, but to point out the diversity of opinion of such endeavours. Byron remained one of the most outspoken critics of the neoclassicist movement in Britain, aligning opposition to the appropriation of ancient artefacts with opposition to imperial programmes. In the midst of national concerns over economy, trade, population and agricultural production, spending £35,000 on marbles was to many an incomprehensible investment, as political satirists were keen to show. One illustration by George Cruikshank, for example, depicted Lord Elgin as John Bull, 'buying stones at the time his numerous family want Bread'.[102] Elgin's speculation made stones more valuable than bread, emphasising the perceived instability of the national economy, confounded by interests in the past taking precedence over concerns for the present.

So in these changing attitudes towards classicism, collecting and museum building, what message might be teased out? That the Elgin marbles were indeed purchased or that Napoleon – even after his defeat – returned to Paris to a triumphal entry festival, is telling of the relationship between politics and art, particularly during periods of political revolution and political turmoil. Moments of social and political crises tend to alter the activities of ordering knowledge and social relations that were taken for granted in both France and Britain. That travel writers and classical scholars developed keen interest in the antiquities that were brought from the ancient lands tells us much about how early nineteenth-century commentators used museums to create historical and imperial narratives. Forming collections for the national estate – for the Louvre or the British Museum – were ways of fashioning a cultural identity that was highly politically charged. On the one hand it seems that the uses of such collections meant whether or not the public could have examples of what was deemed tasteful art and the freedom to enjoy national treasures. On the other hand, the implication of such endeavours could also be turned into nothing less than an endorsement for revolutionary conquest and imperial expansion, as critiques of Napoleon's campaign or the acquisition of the Elgin marbles suggests.

Self-styled caretakers of civility

The Levant represented both an imperial and an historical frontier, where not only the British, but, as was particularly drawn out in this chapter, also the French had interests in penetrating. By considering these competing interests, we see ways that other cultures and ancient civilisations were understood according to the particular ways that British and French commentators chose to represent them. This perspective illustrates for us how artefactual representations of ancient civilisation tell us less about ancient history and customs than about the ways that late eighteenth- and early nineteenth-century commentators conceived of themselves as being civilised. This became the criterion to determine who was the legitimate inheritor of the ancient principles of European rule. This was reinforced by the association made between the civility of the ancients and the self-defined civility of modern imperial rulers, the missionaries for the process of civilising the rest of the world. When exploring this frontier, British travellers refined the notion of a 'European' identity to describe – in nationalistic discourse – who was *most* European – that is, most civilised. This frontier, then, helped to define a British, rather than a shared European, identity. Through travellers' activities, the formation of archaeological collections, and other 'museological' endeavours, both the French and the British articulated ideologies of democracy, liberty and civility. Not surprisingly, such concerns were become most visible during a time of intense warfare and political turmoil.

The moment of social and political crisis during the period under examination altered the activities of ordering knowledge and society which had been taken for granted in both France and Britain. Such altered states render more explicit the concerns at various levels of society over a legitimate and reliable approach to demonstrating knowledge of the history of civilisation. Rapid institutional reorganisation in one nation, coupled with acute political manoeuvring of the other to avoid upheaval, make visible to the historian intense moments of negotiation intent on redefining the cultural landscape. By looking at the development of museum collections and the national and political interests in the objects on display (in places such as the British Museum and the Louvre) during the Napoleonic Wars helps us to understand the values and uses of museological studies in the classification of populations.

Both France and Britain had a long-standing and deeply rooted regard for classicism and accounts of classical civilisation. The newly established French Republic of the 1790s promoted projects to develop the Louvre as a public place where moral messages about virtue, liberty and democracy were to be embodied. However, with the increasing success of Napoleon's military conquests and his gradual control over how the political ideology of the state should be represented, a transformation occurred in the ways that the Louvre was used to express these political messages. This chapter has examined Britain's response to this transition. During Napoleon's reign, an imperial discourse developed which was laden with justifications for collecting antiquities which themselves became emblems of empire. While Napoleon's command pushed back imperial frontiers over Italy, Greece and Egypt, the British government endorsed missions to rescue vestiges of once great civilisations from the hands of the French. Particularly after Napoleon was defeated by the British in Egypt in 1799, the acquisition of collections from the ancient lands for the British Museum became an expression of political and civil triumph over radical and corrupt French tyranny.

We see, then, the creation of a new discourse that was used to express British values and political identity. But the writings and collections of travellers and diplomats to the ancient lands also made fashionable new imagery represented in art and architecture. In the decades following the imperial contests that were surveyed in this chapter, neoclassical buildings were erected as emblems of ancient virtue. For example, William Wilkins' design of the University of London and Downing College, Cambridge, and Robert Smirke's design of the new British Museum in the 1820s; Charles Cockerell's designs for Cambridge University Library in the 1830s, and so on: these buildings represented an ideal of unity of art and science, and the indivisibility of knowledge. Ironically, several writers in the early nineteenth century suggested that Greek revivalism created a new 'modern style'. As Edward Aikin explained in 1808, 'the style of modern architecture is universally admitted to be founded upon what is called the antique.'[103]

While the antique provided a source from which 'modern styles' were created, images and collections of other sorts that travellers brought home were used to assist new forms of pedagogy and to inform new perspectives of European frontiers. The concluding chapter thus looks at some of the things that happen in this regard after travellers return home.

. . . And Back Home

The rare manuscripts and other valuable Books brought to this country by Mr. Cripps, of Sussex, and the Rev. Mr. Clarke, are principally intended to enrich the library of Jesus College, Cambridge. – The Collection formed by these Gentlemen is contained in 183 cases, and perhaps the largest ever sent to England, illustrating the Natural and Moral History of the various Peoples they visited, in a journey from the 69th degree of North latitude to the territories of Circassia, and the shores of the Nile. The Botanic part contains the Herbary of the celebrated Pallas, enriched by the contributions of Linnaeus, and his numerous literary friends. With the minerals, are several new substances, and the rarest productions of Siberian Mines. Among the Antiquities, are various inscriptions and Bas-Reliefs, . . . in the Greek and Latin languages, are several Manuscripts of the Classics, of the Gospels, and the Writings of the earliest fathers of the Church. In addition to these, the Collection contains Greek Vases, Gems, Sculpture, and many remarkable Egyptian Monuments from the ruins of the City of Sais, discovered by those Travellers in the Delta, after the evacuation of Egypt by the French. Also numerous original Drawings, Maps, Charts, Plans, Models, and the Seeds of many rare and useful Plants.

The Times, 8 December 1802, p. 3

5
Coming Home: Displaying and Describing the Trophies of Travels

Collections and displays

When Clarke arrived back in Cambridge, the university members finally saw the face that had already achieved celebrity status in the community. Not long after Clarke and Cripps had arrived in the English port, the *Cambridge Chronicle* happily reported that 'The Lapland Travellers, Messrs. Cripps and Clarke, of Jesus College, are at length safely returned to this country.'[1] What had already safely arrived were the boxes that Clarke had shipped back to Cambridge from the European frontiers, his share totalling about 76 crates.[2]

Clarke and Cripps's collections were noteworthy partly because of the *number* of specimens, and partly because they represented such distant and diverse parts of Europe. But the travellers were by no means unique for having collected during their journey. Museums at the national as well as private level were attracting public notice for the galaxy of gems, marbles, manuscripts and other curiosities that were being hoarded in Britain. The British Museum was fifty years old. By this time, the founding collection of 'Books and Curiosities' that were purchased through Acts of Parliament for nearly £20,000 from the estates of Sir Hans Sloane and Robert and Edward Harley had grown rapidly as a result of successive donations from individuals anxious to contribute to national prestige. The exponential growth in interest in the British Museum was due in large part to the activities of the Trustees appointed by King George II's 1753 Museum Act. This governing body included the highest ranking officials in the British government, as well as the President of the Royal Society and the President of the Royal College of Physicians.[3]

For British ambassadors residing in foreign countries, the British Museum became a natural archive for a wide range of exotic and foreign specimens. The Elgin marbles contributed to an already famous history of collections housed in the nation's museum. Sir William Hamilton, the Ambassador at Naples, achieved national distinction by selling Vesuvian mineral specimens and Greek vases to the Museum for around £8,000 in the 1790s. Such ambassadorial contributions to the nation's museum became the model of conduct for British agents all over the British Empire. Likewise, the agents of the Secretary of State and the Admiralty now had a national vault in which to deposit their exotic payloads. These included curiosities collected on encounters with indigenous peoples during world-wide expeditions, such, for example, as were famously obtained during the voyages of Captain Cook between 1767 and 1779.

The growing scope of collecting by the British over distant lands was facilitated by the mobilisation of naval vessels and government funding. The result went beyond capturing curiosities peculiar to different populations, and extended to the discovery of natural historical specimens that helped define the classification of the kingdoms of nature. Previously unknown specimens – whether from the animal, plant or mineral kingdoms – occasionally presented new characteristics to the eyes of naturalists who refined their classification schemes of known species and specimens. The growing popularity of natural history stimulated searches on foreign shores in the hopes of bringing home new specimens. Naturalists across Europe paraded their treasures around. When Clarke returned from Europe and began to tell others about his collection of natural history specimens, he noted that 'a fine mineral, as well as a fine picture, will often make the tour of Europe; and may be seen in London, Paris, and St Petersburg in the course of the same year.'[4] Likewise, in similar respect to a modern museum, possessing a rare mineral lent prestige to the owner of the cabinet in which it was displayed. In addition to reputation, foreign collections could also provide income; indeed, acquiring exotic curiosities afforded some the currency with which to recoup the expense of travelling. John Henry Heuland, who was part of an extended family of established London mineral dealers, travelled extensively throughout Europe during the first decade of the nineteenth century collecting minerals that supported his business for the rest of his life.[5] A growing market for similar specimens in surrounding shops and touts' tickets for entry to the British Museum reflected the growth of the collections inside.

Natural history specimens collected from foreign shores were prominent investments for the British Museum in the early nineteenth century. In 1809 the MP, Fellow of the Royal Society and natural historian of note, Charles Greville, died. From 1773 he had eagerly collected precious stones and foreign minerals which the trustees of the British Museum deemed 'equal in most, and in many parts superior, to any similar collection which any of us have had an opportunity of viewing in this or other countries'. To acquire it for the Museum, they obtained a special parliamentary grant for the substantial sum of £13,727.[6] The fantastic amounts of money which the Trustees were willing to spend on natural history collections reflects the national investment in organisation and analysis of natural resources around the world for potential economic exploitation. The creation of a skilled workforce within the Museum to organise the collections brought on new concerns about the internal structure of the Museum itself. After the large Greville purchase, Sir Joseph Banks, President of the Royal Society and Museum trustee, suggested that two different displays be constructed, 'the one for the man of science and the other for the stupid gaze of the visiting vulgar'.[7]

Banks's comment draws attention to the problems placed on the organisation of mass education through increasing accessibility to the collections of the 'national estate'. When the British Museum opened its doors to the public in 1759, numerous visitors' comments inform us that the curators neither expected nor were prepared for its incredible popularity among the 'Mechanics and persons of the lower Classes'. Small groups of visitors were allowed entrance only after applying for tickets from the Museum's porter or by obtaining them through external touts. In 1805, after much publicity surrounding the acquisition of Egyptian antiquities and the first round of exotic natural history collections, the ticket system was abandoned. By the time of the Greville purchase in 1810, two years after the first new gallery extending Montagu House (the original building) was opened, it was agreed that all persons 'of decent appearance' would be permitted free entry.[8] Increased visibility to museum collections that illuminated natural and ethnographic history stimulated public discussion and propagated further publications, including guides to a number of other collections. Under the banner of education came a host of arenas for the dissemination of knowledge about foreign territories.

The British Museum was perhaps the most prestigious to house nature's wonders, but it did not stand alone. The Society of Antiquaries continued to accumulate paintings and antiquities from home and abroad, and natural history collections were growing at places such as the Linnean Society (1788), the Geological Society of London (1807) and the Zoological Society of London (1826). Smaller, privately owned collections were also opened to the public. These ranged from odd antiquities and oil paintings collected in the ancient lands to military apparatus or manuscripts. In 1833, the architect Sir John Soane offered his house and collections – everything from architectural models to ancient sarcophagi – to a national Trust for public display after his death (1837). He lamented the government's lack of interest to create modern public galleries to enhance national prestige, a problem partly corrected when Soane's pupil, George Basevi, was later able to design the Fitzwilliam Museum in Cambridge.[9]

In London's Sloane Street, an eccentric collector known as P. Dick opened his 'Museum of Antiquities and Foreign Curiosities', where, alongside thumbscrews and jaw bones, visitors could marvel at East Indian chopsticks, Turkish robes and Lapland sledges and snow boots.[10] In 1819 the traveller, naturalist and antiquarian William Bullock acquired and sold off parts of this collection at auction, along with parts of his own museum. At the beginning of the nineteenth century, Bullock opened his 'Liverpool Museum' which displayed curiosities (from armoury to tattooing instruments) and natural history specimens (from stuffed polar bears to mandarin ducks), which he collected during his own – and acquired after many others' – travels, including artefacts brought back from the South Seas by Captain Cook.[11] In 1812 the museum moved to Piccadilly, and became known as the London Museum, or 'Egyptian Hall', where it remained open until 1819 (shortly before Bullock moved to Mexico). In London the collection was further augmented by donations from the Lichfield Museum, and the selling off of the Leverian Museum in Leicester Square (the natural history collection of Sir Ashton Lever).[12]

These were just a few of the many public and private galleries that were opened for the public gaze in the early nineteenth century. Through these places the material underpinnings of foreign societies were presented to the public. Collectors of the goods themselves sometimes reached celebrity status. Clarke certainly turned heads when he returned to Cambridge and began displaying and distributing parts of his collection of European curiosities.

Even in this local context, the extent to which Clarke's collection impressed members and visitors to Cambridge cannot be understated. Some twenty years after returning, an historian of Cambridge emphasised that Clarke and Cripps had 'done themselves and their country so much honour by their zeal and perseverance in research, during their very extensive travels', and added that it was to their honour that they were able to bring back such a magnificent collection, things that only 'have been described by different travellers'.[13] Placed in the vestibule of the Cambridge University Library was the most conspicuous treasure from Clarke's collection: the two-tonne statue of Ceres. Shortly after, the last relic relating to his journey was placed not far, near the entrance to the east room of the Library, set in a large mahogany case. Straight from the Society of Antiquaries in London came 'a beautiful facsimile . . . of the remarkable triple inscription found at Rosetta', the stone that Clarke had helped to procure from the French.[14] It would probably be some time before a Cambridge traveller exceeded Clarke's service, but not for lack of interest. As Clarke proposed (in a self-satisfied account of his own achievements) in the Preface to his description of his *Greek Marbles brought from the shores of the Euxine*:

> If future travellers from the University, hereafter visiting the territories in which these monuments were found, contribute also their portion, Alma Mater will have no reason to blush for her poverty in documents so materially affecting the utility and dignity of her establishment. The foundation, at least, of a collection of Greek marbles may be said to have been laid . . . and some points of antient history may appear illustrated.[15]

The University Library was the 'museum' in Cambridge until the Fitzwilliam Museum was completed in the 1840s. Clarke's collection joined what was already an impressive display, with antiquities donated by a number of eighteenth-century travellers. But in addition to the Library, Clarke dispersed a significant proportion of his possessions to friends and other institutions, an act which at least one current authority believes marked 'the beginnings of a more philanthropic spirit in the acquisition of antiquities', rather than collecting merely for private enjoyment.[16] The majority of his ancient manuscripts were divided between Richard Porson, the Professor of

Greek in Cambridge, and Oxford's Bodleian Library. The latter paid £1,000 for 45 volumes of oriental manuscripts, and 40 Greek and Latin manuscripts, including a volume of dialogues of Plato, written by John the Calligrapher in 895, one of the present treasures in the Library's collection.[17] He gave the British Museum a range of gems, vases and a tomb proclaimed to be the resting place of Alexander the Great. Most of the remaining objects Clarke kept for himself but, between his donations, sales and publications, his reputation as a traveller and collector already shone across England.

His publications relating to his travels and his collection were completed with the assistance of well-known *literati*, most of whom had recently completed their own tours. The publication of Clarke's books describing Ceres and the tomb of Alexander in 1802 and 1805, respectively, were written with assistance of the famous satirist and Italian scholar Thomas James Mathias and the distinguished artist and sculptor John Flaxman, whose engravings illustrated the works. [18] A few years later, when he was preparing the first volume of his *Travels* for publication, he borrowed illustrations from George Hamilton-Gordon (Lord Aberdeen), who had returned from his own travels in Greece in 1803 and had founded the Athenian Society. Sir William Gell, Clarke's Jesus College friend as well as contemporary traveller to Greece, corresponded with him about their travels, and Gell went on to establish himself as a renowned classical scholar and topographer.[19]

Clarke's collection was held in high regard in an age when museum collections were increasingly being transformed from amusement areas to educational centres. Travellers' collections were trophies of travels – they provided an opportunity for others to engage with material culture of foreign places. But complementing this was the dazzling world written into descriptive texts, either as guides to collections or narratives of the traveller's journey. Brandishing their treasures boosted travellers' reputations, but the most widespread generator of reputations for literary travellers was the published account.

Writing travels

For over a decade after his return the public waited to see the first linear narrative of Clarke's journey to the European frontiers. 'The state of the world forbids any sanguine expectations that an opportunity will soon be afforded of carrying on new journeys,' wrote

Henry Brougham in 1808, taking comfort that at least previous travellers' narratives were still being published. 'But why does Dr E.D. Clarke delay to fulfil obligations which he long ago came under to the literary world?'[20] Finally, in 1810, Brougham's longing was satisfied. Having seized the opportunity to pore over the first volume, he warmly welcomed Clarke's contribution to the annals of travel in an article for the *Edinburgh Review*. Brougham thought Clarke deserving of praise for having made 'a long and laborious progress through countries little visited, or much misrepresented by other travellers: he has had the enterprize to encounter both hardships and dangers in the pursuit of useful and interesting knowledge; – he has plainly and sensibly related his adventures; – he has observed carefully, and often wisely; – his learning on some subjects, as botany and antiquities, is minute and copious.' In sum, Clarke's indefatigable efforts as a traveller and writer were useful 'for increasing our knowledge of countries scarcely civilised, but yet aspiring to the first rank of European nations; – and for introducing us to an acquaintance with tribes scarcely at all described by preceding travellers.'[21]

Clarke's first volume was well received and his timing was right. In the 1810s, more travellers to the Continent – particularly amongst the Cambridge group who had colonised Athens in the previous decade – were writing up and publishing accounts of their travels. The publishing trade was beginning to sizzle with continental travelogues and authors were becoming more competitive. One admirer of Clarke's first volume who was in the process of penning his own reminiscences of travel was Lord Byron. He met Clarke in 1811 when the former Trinity College student returned on a brief visit to Cambridge to witness the inauguration of the new Chancellor of the University.[22] At the time, Byron was also looking forward to the moment when his travel companion, John Cam Hobhouse, would produce his own travelogue of their trip to Greece the previous year. But he also knew that Clarke was at work on his second volume, which was likewise to be devoted to Greece. So was Hobhouse, who was anxious that his own travel book be published before Clarke's second volume, which promised to be just as popular as the first. While preparing his *Journey*, Hobhouse wrote to Byron: 'Clarke's travels, vol. 1st from St Petersburgh to Constantinople, are come out; they are reviewed both in Edinb. and Quarterly; rather more favourably in the first than the last, but in both favourably: indeed, they seem to be most excellent. His next vol. is to take in

Turkey and Greece, which will render any other book on that subject quite superfluous.'[23]

In light of the competition for selling new information and hitting the market before it became exhausted, Byron gently probed Clarke about the contents of his second volume. More conspicuously, Byron was aware that Hobhouse intended to append an exposition on Romaic (modern Greek language) to his account of Greece, and wondered what Clarke had to say on the matter. Acting as intermediary between Hobhouse and Clarke, Byron was able to allay Hobhouse's fears of competition. 'We are to dine at Dr. Clarke's on Thursday,' he wrote to Hobhouse. 'I find he knows little of Romaic, so we have *that* department entirely to ourselves. I tell you that you need not fear any competitor, particularly so formidable a one as Dr. Clarke would probably have been.'[24]

It was also through the encouragement of Hobhouse's publisher, James Cawthorn, that Hobhouse pressed on with his work, but the correspondence continued with worry. Hobhouse frantically wrote that his '*Travels* are going on swimmingly – plain prose is to be my fate – you shall be immortalised you rogue you shall. . . . Clarke's *Greece* will not be out for 9 months – if I can but cut in before him!!!'[25] Byron once again comforted Hobhouse that 'I do take much interest in [your Quarto] & have no doubt of its success.'[26] As Byron noted, he was in a position to assure Hobhouse that his volume would succeed because the content of the book would illuminate an area of Greek culture that Clarke would not address: modern language. Indeed, even though Clarke did publish his volume on Greece (1812) before Hobhouse (1813), both were received well – an indication of the growing public interest in that region of the world.[27]

The more travel accounts that were published, the more they whetted the public's appetite for information about distant lands. Publishers and booksellers looked favourably upon new travel books, and reviewers further integrated new nuggets of information about foreign countries with daily discussion of current affairs. For readers of any one of the numerous popular reviews that poured off the presses and into their postboxes in the early nineteenth century, everything from literary gossip to economic forecasting, from warfare to fishing, was served up with breakfast. The broad coverage of news and events (foreign and national) written in the pages of weekly periodicals or monthly magazines created what is now referred to as the 'common context' of social, political and scientific thought.[28]

Periodicals propagated more easily new observations and opinions, but they also acted as vehicles for advertising new travel books (in addition to other forms of literary genre).[29] Travel narratives were in this way not only woven into the matrix of intellectual discussion, but in this medium they were reduced to basic units of information and easily related to different discussions of current events.

Throughout the eighteenth century gentlemen were advised that travel literature provided 'the chief materials to furnish out a library', according to the third Earl of Shaftesbury, and their popularity soared.[30] The competitive marketplace yielded high rewards and prestige for successful authors. No good library was without its edition(s) of Clarke or Coxe; political tracts such as Malthus's *Essay on population* (packed with data accumulated through travel), or books that promoted agricultural or social improvement, such as Arthur Young's *Travels . . . with a View of Ascertaining the Cultivation, Wealth, Resources, and National Prosperity of the Kingdom of France* (1792). These books not only further supported the centrality of travel in multiple social discourses, but also helped make travel a precondition for enhanced literary and social reputation. The fashion of interspersing writings and speeches with 'travell'd learning' was even picked up in political satire – such as John Hawkins' jest against the MP Sir Nathaniel Wraxall.[31]

While the market sustained individual, lavishly produced travelogues, it also accommodated compilations and summary collections assembled from the narratives of different travellers. These wideranging works of synthesis brought together such material as statistics on foreign population, chief items of import and export, and observations on trade and manufactures. A good example of the reproduction of raw data from a variety of travellers' reports is John Oddy's *European Commerce*, which compiled facts and figures on trade and commerce relating to the northern countries including Sweden, Russia and Germany.[32] Compilations helped make travellers' accounts, and the variety of data within them, more easily accessible.

The rationale for publishing collected and abridged travelogues was in part to draw out and make even more explicit comparisons between various travellers who tended to have a predilection for a particular branch of study. Travellers, taken in isolation, tended to pay too strict attention to specific objects. Of course, whether their attention be directed to physiological or antiquarian research, natural history or to military matters, each traveller's labour was usually considered deserving of public approbation and encouragement. But the efforts of travellers *in toto* resulted in an unwieldy sea of knowledge through which readers had to wade in an attempt to broaden their horizons, familiarise themselves with their neighbours and rethink their prejudicial attitudes. Francis Blagdon, the editor of a collection of travels titled *Modern Discoveries*, explained the relevance of travelogues:

> It is universally admitted, that no books are so completely calculated to afford instruction and entertainment, from the magnitude of the objects they embrace, as those of Travels in distant climates; for by the perusal of their general contents, independently of the portion devoted to scientific inquiries, a contrast is formed with the habits and customs of our native land; and thus the juvenile mind, instead of becoming contracted and prejudiced, is accustomed to reflect with advantage on the different shades of human nature, and to reconcile its manners in various parts of the world.[33]

In order to help convey lessons about the 'different shades of human nature', he prepared cheap, widely accessible anthologies of travellers accounts. The aim of *Modern Discoveries* was to take information from the elegant quarto volumes stocked in 'splendid and expensive libraries only' and make it available 'at a price which can neither be felt nor regretted by any enlightened individual, whatever may be his rank in society.' Thus, Blagdon's promise to the public was that his volumes would be printed in 'Pocket-size' on 'fine woven paper'. One volume a month would be produced, sold 'at the very moderate price of 5s'.[34]

Some editors drew from authors that would feed a particular theme, such as the illustration of European geography, which in the early nineteenth century grew increasingly sophisticated from the efforts of travellers such as those discussed in earlier chapters of this book. One editor of this sort was the Scottish antiquary and historian John Pinkerton. His *Modern Geography* was first published by Cadell and Davies in 1802, but went through numerous editions and further abridgements over the following couple of decades. The first volume of this work, in which he extracted 'the essence of innumerable books of travels and voyages', was devoted to European geography. Here Pinkerton demarcated states in a variety of ways, first into different orders according to different states' 'real consequence' and then according to different political and civil criteria, as well as physical geography.[35] It was an ambitious project, but at least one reviewer in Cadell and Davies's *Imperial Review* reckoned it was useful for illustrations of 'general facts' about European geography![36]

Sir Richard Phillips, a radical London bookseller and publisher, also published (under the pseudonym 'J. Goldsmith') geographical and scientific works that were compilations from numerous travellers' narratives. In his *General View of the Manners, Customs, and Curiosities of Different Nations*, which was in large part a kind of geography class-book in which he emphasised the need for students to read and draw maps, he culled information from a bibliography of over seventy travel books that covered America to China, and Iceland to India. His European chapters drew from many of the authors discussed in this study, including Tooke and Bell on Russia, Chandler's *Travels in Asia Minor*, Swinton and Coxe on Scandinavia, and Consett's *Tour of Lapland*. Phillips introduced the geography of the world by tracing the boundaries of Europe, adding observations on the characteristics of the inhabitants of the frontiers. The extreme northern boundary was Lapland, and summarising travellers' observations, Phillips was able to write that the 'Laplanders are hospitable, generous, and courageous. They make long excursions upon the snow, and will, without much fatigue, travel fifty miles a day; and in their sledges, drawn by reindeer, they pass over hill and dale, two or three hundred miles in a day.' The Russian empire was succinctly summarised as 'the largest in extent in the world, comprehend[ing] all the northern parts of Europe and Asia, but only a small part of its inhabitants are in a state of civilisation.' The southern frontier comprised 'Turkey in Europe', which 'includes ancient Greece and other countries, formerly the finest in the world, but owing to the despotism and wretched policy of the Turks, now the most desolate and miserable.'[37]

Travellers' reports were used not only for political or geographical information, but, as further examples from Phillips's compilation illustrate, also to provide evidence for philosophical theories. In this respect, we see Phillips concerned to compile from travellers' reports information about different ways of life, to correlate geographical conditions with customs and – donning the hat of an anthropologist – to chart the differentiation of races. Launching his discussion of the classification of different populations of the earth, he used three 'principal varieties discernible among them' to distinguish '1st their colour; 2d. their figure and stature; and 3d.

the dispositions of each different people'. The climate in which different populations resided was worth particular note since customs and character seemed to depend on different physical conditions of life. Thus, 'all the Tartars, for example, are tawny, while the Europeans who live under the same latitude are white. The difference may be safely ascribed to the Tartars being always exposed to the open air; to their having no cities and fixed habitations; to their sleeping constantly on the ground; and to their rough and savage manner of living.' When the environmental conditions are reversed from what the Tartars experienced, such as the extreme cold that the Laplanders lived in, the physical effect was very similar. 'Both cold and heat dry the skin, and give it that tawny hue which we find so much in so many different nations. Cold contracts all the productions of nature; the Laplanders, accordingly, who are perpetually exposed to the rigours of frost, are the smallest of the human species.'[38]

Phillips thus drew on numerous travellers' reports about the appearance of populations at the 'fringes' and 'extremes' of the 'most temperate climate' of central Europe – which, of course, figured as being the most civilised. He used this empirical data as evidence to support his argument for monogenesis (the attempt to trace European national genealogies to a descendant of Noah).

> Every circumstance concurs in proving that mankind are not composed of species essentially different from each other; but that on the contrary, there was but originally but one species, who, after multiplying and spreading over the whole surface of the earth, have undergone various changes by the influence of climate, food, mode of living, epidemic diseases, and the mixture of dissimilar individuals; that, at first, these changes were not so conspicuous, and produced only individual varieties, which, afterwards, became specific, because they were rendered more general, more strongly marked, and more permanent, by the continual action of the same causes; and that they have been transmitted from generation to generation, as deformities or diseases pass from parents to children.

Support for monogenesis could be used in arguments to suggest that, if all species were reducible to one stock, then racial differences were not providential, but matters of certain external environmental conditions. Evolutionary theorists were often radical political philosophers, as was Phillips, who was known for his strong republican views, and his bookshop even became a depot for radical revolutionary literature. No God-given hierarchy, but people subject to regeneration and degeneration depending on where in the map of the world they lived. Hence apparently similar races at geographically distant (but similarly environmentally harsh) parts. 'In Lapland, and on the northern coast of Tartary, a race of men are met with, of an uncouth figure, and small stature. Their countenances are equally savage as their manners. They appeared to be a degenerated species, are very numerous, and occupy vast regions.'[39]

Phillips' society still secured its place as the top of the civilising scale. But it was so not because of divine endowment, but because of the environmental context was most favourable in central-western Europe. At least, travellers' reports provided information about populations that could be used in support of this theory, as well as other sorts of 'raw data' with which the British began to piece together pictures of the world around them.

Travellers' reports were also used as eyewitness accounts to compare with historical descriptions or legends. Relative to the rise of biblical criticism in the eighteenth century, each traveller to the Levant carried potentially potent views. Travels to the land where the scriptures were written would illuminate manners and customs of Eastern peoples, observations thought especially useful in fully understanding the Bible. Particularly relevant were contemporary challenges to biblical exegesis and the construction of the 'myth of Christianity'.[40] At issue here was the historicity of Christian revelation. Recent scholarship has shown that the 'received interpretation' of Scripture, as adumbrated in the late eighteenth century by the approximately 230 books on religion that were published each year, relied on knowledge of the language, geography and customs of the East.[41] Challenges to traditional biblical interpretation came partly in the shift in emphasis from literary comparisons of the Bible with Greek and Roman classics, to an emphasis on topographical, historical and empirical investigation of the text.[42] Travellers' accounts worked to prove, by witnessing the material, ancient evidence held against textual variations, the authenticity of the Word. Consequently, the traveller proclaimed himself an authority on the basis of eyewitness testimony. When Clarke visited the Holy Land on a sojourn from Greece in 1801, his biographer tells us that 'the Bible was in his hand, and the book of nature lay open before him.' The same writer received an enthusiastic letter from Clarke:

Jerusalem, 10 July 1801. The date! – the date's the thing! You will thank me for a letter dated *Jerusalem*, more for the little local honour stuck in its front, than for all the fine composition and intelligence it may contain. . . . To me it appears as though the eyes of former travellers have been entirely shut upon their coming here; or that they were so occupied by the monks and their stories, that they neglected to go out of the walls. To those interested in evangelical history, no spectacle can be more mortifying than Jerusalem in its present state. The mistaken zeal of early Christians in their attempt to preserve has, for the most part, annihilated those testimonies, which might have remained at this day to establish the authenticity of the Gospel.[43]

In a way reminiscent of his revaluation of Swedish natural historians' accomplishments compared to the 'enlightened' travellers from Cambridge such as himself (discussed in chapter 2), Clarke here reaffirmed his merits as a literary traveller (comments which resonate in his published *Travels*) by proclaiming that 'the eyes of former travellers have been entirely shut.' Clarke's role as a traveller was being crafted to complement an important social image of a traveller-as-authority, an image especially propounded in Cambridge, where 'those interested in evangelical history' abounded.[44] As we will see below, Clarke made further use of his travel narratives to illuminate ancient texts in his Cambridge lectures.

But others also relied on travellers' observations to supplement textual criticism. In *The Eastern Mirror*, W. Fowler informed his readers that the 'valuable works of Mess. Harmer, Calmet, and Bruder, upon the subject of Eastern customs, are too expensive to be acquired by some, and too voluminous to be read by others.' His compilation was published to remedy these inconveniences, and he was able to add observations from a number of other travellers to the East, including Shaw, Pocock, Chandler, and Clarke. The writings of these authors, wrote Fowler, suggest how 'judicious travel' helped add new laurels to the 'Christian system', while 'publishing accessible compilations' helps 'in perfecting the Christian character'.[45] In his book, passages from the Bible were reproduced, with excerpts from travellers' reports illuminating literary allusions. Descriptions of contemporary customs were invoked to illustrate ancient references. For example, to help explain what was fully meant by 'Blessed shall be thy basket and thy store' (Deuteronomy, xxviii, 5), Fowler gathered an explanation from Clarke and Harmer. 'As the *basket* was used to collect their fruits,' the explanation went, 'the *blessing* of the basket may allude to the olive-gathering and vintage. The *store* signifies their leather bags, in which they used to carry things in travelling. Understood after this manner, the passage promises Israel success in their commerce, as the next verse promises them personal safety in their going out and in their return.'[46]

It was part of the function of travellers' written reports to create a dimension of literary criticism where readers could learn about different customs. But by the early nineteenth century, travellers' observations and narratives had also become part of new forms of pedagogy that introduced new world-views to students of a liberal education. It is to an example – based in early nineteenth-century Cambridge University, through the efforts of Clarke – of how travel narratives became part of classroom instruction that we now turn.

Travel and new forms of education

Clarke's Cambridge career after returning from his continental journey provides a good example of how travel narratives could be used to develop new forms of pedagogical discourse. Soon after his return, and having resumed residence in Cambridge and distributed his gifts, Clarke began acquiring academic prestige. In 1803, the University Senate passed a grace (a decree) to award Clarke an LL.D. degree and Cripps an MA. Following this, the Master of Caius College, Martin Davy, was elected Vice-Chancellor of the University. In his inaugural address, he was 'singing out Clarke's praises in the Senate House', emphasising 'the credit which had accrued to the University from [Clarke and Cripps's] labours and public spirit'.[47] Davy, who himself had travelled for three years between 1796 and 1798, and resided in Italy during the French occupation, was a staunch Whig, a noted liberal regarding University politics and an outspoken critic of William Pitt as prime minister and parliamentary representative for Cambridge.[48] As Davy was inaugurated during the Peace of Amiens, he was no doubt particularly elated about the apparent end of the war, the peace with France and had written his speech with a mind to encourage future travel amongst Cambridge graduates.

About three years later, in 1806, following more donations and publications by Clarke, the new Vice-Chancellor of the University and Master of Jesus College, William Pearce, granted permission for Clarke to display the remainder of his minerals, marbles and oil paintings to the public in a lecture room in the Botanic Gardens. This provided a context in which Clarke was able to retell his traveller's tales. This was indeed a privileged site for him and a new opportunity for him to become more visible among the students.

Since his return, Clarke had been meeting Jesus College under-graduates and 'instructing' them on 'rules for travellers'.[49] But, like many 'courses' offered by college fellows, which were usually 'cram courses' on subjects covered in the end-of-year examination, Clarke most likely only met members of his own college. 'Public' lectures, on the other hand, were open to all members of the University. For these, advertisements or lecture notices would be printed, at the instructor's expense, and posted in various University buildings.[50] If interested in attending the course, which would usually last one academic year, a student could register at the local bookseller. Students would pay the average fee of 5 guineas for the year, and, if available, obtain a course syllabus for about £2. The cost of the course was at the instructor's discretion; he would pocket the profits. For 'public' lectures to be offered by non-University professors – even lectures as extracurricular as Clarke's travel narratives – permission from the Vice-Chancellor was needed. This was what Clarke obtained from Pearce in 1806. The timing was appropriate: that year saw Lord Elgin's safe return from imprisonment in Paris, and the prepa-rations for the opening of his Mayfair mansion in London to a select public to view the marbles, which stimulated further interest in foreign travel and collecting.

Clarke's lecture room in the Botanic Garden was set up as a daz-zling display of artefacts from exotic locations, an obvious allurement to undergraduates. The day after his first public performance, Clarke enthusiastically wrote to his close friend William Otter to report his popularity and success:

> I have only time to say, I never came off with such flying colours in all my life. I quitted my papers and spoke extempore. There was not room for them all to sit. Above two hundred persons were in the room. I worked myself into a passion with the sub-ject, and so all my terror vanished. I wish you could have seen the table covered with beautiful models for the Lecture.[51]

His collections set the stage and accompanied his narratives. The models were displayed on a large table which almost spanned the width of the room. They included wooden models of minerals showing their fundamental crystallographic forms, which he acquired in Paris (a stop on their return from the Levant) while studying with the renowned crystallographer René Just-Haüy. Next to the table sat a large cork model of Mount Vesuvius which Clarke had constructed while in Naples (earlier in the 1790s) after making a dozen trips up the side of the mountain. Sir William Hamilton, the British Ambassador to Naples, even praised it as the best model he had ever seen.[52] Clarke's model was especially interesting since it could be made to erupt lava, like the real one as depicted in the oil paintings which hung on the walls around the lecture theatre – further props for Clarke to frame his lectures. The cabinets were designed and situated to display the exotic minerals he collected from around the world, placed with reference to their different geographical origins (a common system of arrangement throughout the eighteenth century). All this he synthesised into the content of his lectures.

Clarke advertised that he would discuss the 'specimens exhibited which had been collected by the Professor during his travels'. His lectures were designed 'to illustrate the natural history of the materials used by Architects, Sculptors, and Lapidaries, in the remotest periods, and in modern times'; and to illustrate 'the Mineralogy . . . of the ancient Poets and Historians'.[53] To the students, Clarke was presenting material products from a world that most of them had only experienced through texts. Contemporary accounts from both Clarke and his students provide lucid descriptions of the lectures.

Clarke's 'delivery was a master-piece of didactic eloquence', recollected one former student. 'Even the commands to his attendants were, somehow or other, squeezed into the sentence as to produce no interruptions in the flow of his discourse. . . . From every stone, as he handled it and described its qualities – from the diamond, through a world of crystals, quartz, lime-stones, granites, &c. down to the common pebble which the boys pelt with in the streets, would spring some pieces of pleasantry.'[54] Clarke described the natural history of each mineral, combining narration of where he had collected them, an account of how natural philosophers classified the mineral, as well as reflections of how ancient authors had described them. The combination of these themes enabled him to situate his lectures within the historical context of ancient myth and literary reference. Another student recollected that

> Clarke, the enthusiastic traveller . . . [filled] the lecture room [with] beautiful specimens, which are so delicately arranged upon the

table, and the surrounding cases, from the primitive formation of granite to the costly stores and precious metals; . . . the picture of the grotto of Antiparos, with its beautiful stalactites and crystal floor; the igneous section of the strata of the island; the green god of the New Zealanders; and a vast collection of curious and precious things.[55]

No longer would students simply read about foreign or classical civilisation, now they could be empirically stimulated directly by the artefacts of the world described.

After attracting large audiences for nearly two years, new negotiations took place which were critical for Clarke's career and the structure of the Cambridge curriculum. In 1808, after persistent canvassing by college fellows and friends, the University's governing body passed a special grace awarding Clarke the official title as 'Professor of Mineralogy', – the list of professors now totalling twenty-three.[56]

Designating him Professor of 'Mineralogy' might seem rather curious. Why place him in the sciences, and not make him professor of arts? Or travel? But it is necessary to remember that disciplinary boundaries such as our modern conception of the cultural gulf between the arts and the sciences did not exist in the early nineteenth century. Further, 'mineralogy' did not refer to a precise scientific pursuit; it could refer to practices we now associate with geology, chemistry or antiquarianism, or general natural historical interests in travel and collecting specimens. Broadly speaking – and these were the terms on which Clarke's professorship was conceived – 'mineralogy' was concerned with pretty much anything that came from the earth: natural history specimens as well as culturally crafted artefacts. But in a similar fashion to travel accounts themselves, with a function to educate about distant lands and different people, Clarke's collection of curiosities afforded another way to pass on knowledge about the world. It was his opinion, and precisely his objective in establishing a course of lectures on mineralogy, that to educate pupils effectively about the world was to use visual aids to illustrate the different cultural and historical uses of minerals and other resources of the earth.

The Cambridge fellows, among whom the governing body of the University was formed, saw Clarke as achieving something more than *éclat* as an eccentric traveller. His engagement with the classics provided a fresh way of analysing literary discourse: he provided new pedagogical resources with which to inspect classical texts, antiquarian artefacts and foreign customs. He offered new perspectives on 'literary criticism', which during this time was broadly defined.

The editors of *Museum Criticum; or, Cambridge Classical Researches* welcomed any contributions to their publication that were 'original communications, such . . . as may contribute to throw any new light on the manners and customs, arts and sciences, history and antiquities of the Greek and Roman empires, whether from the observations of modern travellers, or the stores of ancient learning.' In short, the object of literary criticism was to pay 'critical attention to the languages of antiquity'.[57] This Clarke certainly did, and his efforts were rewarded not only by acknowledgement in the journal and personal support from the classicists, but in the procurement of his more prestigious position as professor.

Clarke's good friend, principal drinking companion and Professor of Greek at Cambridge until his death in 1808, Richard Porson, had expressed what he believed to be the best possible basis for 'a sound and liberal education'. Classical studies, he eloquently stated, should not 'be stigmatised as the mere study of words, to the disregard of things; for as words are the signs of things, no one can think of words without being led at the same time to think of things.'[58] The problem was, words could be misleading. It was to Clarke's credit that he could refer directly to the object under discussion. 'Our knowledge of antiquity is drawn from two sources, – monuments and antient authors,' explained the arch-pundit of the antique, Richard Payne Knight, in a review of Clarke's treatise on the Greek marbles he shipped home during his travels. 'The latter, though far more copious, can never be so decisive as the former, both on account of the corruption of manuscripts, and the difficulty of representing to our minds images of things which we have never seen.'[59] Literary critics from biblical scholars and classicists to travel writers grappled with the problems of textual representation.

In his published *Travels*, Clarke explained to the reader 'the importance of attending to every object likely to serve as a *land-mark* in the *topography* and *geography* of Greece', and of 'any other apparently trivial relique connected with the antient history of the country; – not being aware, that, in very many cases, these remains are the only beacons we can have, to guide our course, in penetrating the thick darkness now covering this "land of lost Gods and men."'[60] Clarke was writing a *guide*; on the one hand, a literal guide with which future travellers could plan their route. On the other hand, it was a guide 'in adapting passages from antient authors for the illustration of its antiquities and history'.[61] The account of ancient lands by classical authors was continually being checked against the observations of modern travellers. In this way, adaptations – corrections and clarifications, or just verifications – were made to ancient (and modern) textual accounts of the ancient lands. Given that works such as Pliny's *Natural History* were such heavily tapped resources for eighteenth- and nineteenth-century commentators, the ability to contrast passages with current knowledge could merit critical literary acclaim.[62] In his lectures, Clarke continued to make connections between different historical, cultural and intellectual worlds.

With his model of Vesuvius and the chemical instruments he learned to use while touring the Swedish mining districts, he combined lessons on modern chemical and mineralogical systems with an historical narrative of the history and progress of civilisation. This was accomplished, for example, by chemically analysing ancient vases to determine the materials used in their manufacture. Noting changes of the different types of clay used in different historical periods indicated, he explained, a maturing understanding of the most appropriate materials for the production of pots, vases, and the like. Similarly, in a lengthy discussion of the different varieties of marble, students would learn how it was used among the ancients – for example, what kind of marble was used in the construction of the Greek Parthenon and statues.

Henry Warburton, a student who attended Clarke's lectures between 1811 and 1814, recorded in his lecture notes that 'the ancients used the Parian and Pentelican Marbles – the latter whiter and much more beautiful – used by them in all their best Statues – but from the prevalence of veins of extraneous matter, extremely liable (in consequence of their decomposition) to flake to pieces.' For a visualisation of this, the student needed only to visit the vestibule of the University Library where Clarke donated the colossal statue of Ceres which he brought back from Greece. 'The mutilation of the face of the Elusinian Ceres', Warburton went on to note, was 'the consequence of veins of Schistus . . . Hence nearly all the best statues of the ancients being form'd of this marble have perish'd, and those only of Parian, a purer marble, but not so white, have remained.' While this emphasised the way different countries at different times transformed earth to art, Clarke also made regular reference to literary texts and pointed out what minerals were being alluded to.

In his discussion of emerald, for example, Clarke suggested that 'Pliny and Theophrastus mention pillars obelisks &c made out of Emerald' but, due to the geographical area referred to, the authors 'probably allude to the green Horn Stone'. *'Emerald'*, he went on to explain, was 'a name they gave to any greenish transparent stone.'[63] In an example of the different customs surrounding the uses of minerals, he made reference to the biblical account of Jezebel painting her face. 'And when Jehu was come to Jezreel, Jezebel heard of it, and she painted her face, and tired her head. . . .'[64] He then told a story about how the 'women of Circassia, Georgia and the Levantine countries [used antimony] as a pigment for the eyes.' Warburton learned that painting her face 'is literally in the Hebrew painted her *eyes* and has reference to this custom'.[65] The reason the English translation of the Bible was so poor, written by people who had apparently not witnessed this custom for themselves, was because the translators were 'unable to reconcile with their notions of a female toilet'.[66]

In this manner Clarke used his artefacts and literature both to extend his geographical accounts of the lands he travelled and to illustrate his stories about the customs of people in distant countries. When Clarke advertised that in his lectures he would talk about the 'Mineralogy of the ancient Poets and Historians', he meant applying modern mineralogical and chemical analysis to the marbles, about which the ancients wrote. He broadened the perspective of literary criticism in Cambridge by further arguing that to understand classical literature and civilisation required the application of his mineralogy. In other examples, we could see how other parts of his collection – from Russian maps to Lapland garbs – could also be used to extend his students' comprehension of foreign territory and customs.

Readers of his travel books as well as students in his Cambridge lecture room were taken through exotic goods from distant cultures. He displayed precious jewels from Turkey; carved jasper-agate vases; distributed (empty!) Turkish liquor bottles to colleagues; deposited manuscripts and copies of the inscriptions of ancient ruins in the University Library; showed off a number of statues, granite slabs with inscriptions, Lapland musical instruments, mechanical locks and keys; and donated sacks of crystals, minerals and stones to the Woodwardian geological collection. Planted in the university botanical garden were seeds gathered from the European frontiers.[67] Drawing on his collections and his experiences of travelling enabled him to spin out endless tales and depictions of historical, geographical and cultural dimensions of life around Europe. The lessons were further refined with the use of literary and scientific tools which in the early decades of the nineteenth century became increasingly part of educational programmes. Through the example of Clarke, we see how in a number of ways foreign frontiers were increasingly becoming part of the purview of a liberal education.

Such approaches to establishing new forms of knowledge about Europe seduced others into dreaming about making 'literary' or 'scientific' travel part and parcel of their own education. Simply embarking on a tour to the Mediterranean was old-fashioned and insufficient for the demands of understanding modern society. What was once for the patrician elite and their hired Oxbridge tutors was increasingly made accessible to others through ease of access to Europe. Certainly by the 1820s, the increasing exposure of the British to the Continent, through travel and publishing, created a condition which many believed would enable modern society to be governed better. 'Much of that learning which was before attainable only by those who studies the dead languages, is now placed within the reach of those of moderate fortune,' wrote the educational writer J.S. Walker. 'And the works of the British historian, the philosopher, the traveller, the mechanist, and the essayist, supply a stream of information which elevates our country to the distinction of a classical and scientific land.'[68] But increasing stocks of knowledge also required refined strategies for acquiring new forms of information. Disciplined travellers needed preparation.[69] To continue briefly with another case from Cambridge, consider the vision of a Trinity College student, Charles Kelsall, in the 1810s.

Kelsall was an outspoken Whig, an advocate of democratic and educational reform, and the son of a high-ranking member of the East India Company. In 1814 he published a *Phantasm of an University*, containing an extravagant proposal to erect a new, ideal university. In the book he considered that changes in curricular structure and architectural design would produce a 'culture of understanding' in which 'the student's career in the science or art for which he shows genius or inclination is best facilitated'.[70]

The problems that he felt plagued Oxford and Cambridge were made explicit. Their curricula were too narrowly concerned with abstract exercises in mathematics, the 'Pure Analytics'. Verbal criticism and 'philological controversy' were paid too much attention. 'Another crying defect in the method of prosecuting studies in either University', he continued, 'is the want of due attention to the theories of Agriculture, Commerce, and manufactures; the bulwarks of the nation, the pride and glory of the English people.' Encouragement to study the political sciences was lacking and, in consequence, 'promoting the public advantage and happiness' was not cultivated. Too little attention was paid to moral philosophy, which he believed was evidence of 'a strange prejudice . . . that inquiries into moral truth tend to sap the foundations of religion.'[71] All of these led to the ultimate criticism that the limitations of the existing university structure did not allow the freedom of individual academic improvement.

Unlike the particular administration of the Cambridge Senate House examinations, a university, he asserted, 'should be able to face, and confer rewards on every candidate in every department of science and art; she should be to a nation what the sun is to our system, the grand centre, from which the rays of universal knowledge should emanate, and by which the career of all the luminaries of science should be regulated and directed.' The more subjects available, the more opportunity to nurture genius. If students' minds were allowed to choose and pursue one rational subject, rather than a multitude 'crammed' into an examination, their intellectual faculties could flourish. As a consequence, 'Insanity, that dire scourge of society, would, in all probability, be less frequent.'[72]

After sounding his criticisms against traditional education, Kelsall projected a probable future for graduates of his ideal institution. He created the lives of seven young men together with a biographical sketch of their backgrounds. In this story, they met at the university. Upon their graduation, one of the students gave a speech to his friends. 'But wherefore all this toil and trouble, this constant trimming of the midnight lamp, if we do not turn our knowledge to useful and brilliant account?' asked Thraso, one of the wealthier students. 'Is it to be forgotten at Newmarket, or evaporated in the fox-chase?' He thought not. 'Foreign travel appears to me to present the most rational and pleasant method of filling up our time; and we will persuade our mammas to spare us for the next three years.' With fresh knowledge from their university pursuits, they should set off immediately. 'But, instead of wearing ourselves to death by each pursuing a multiplicity of objects, we will *divide* our labours.' The division of labour espoused here would be a continuation of the student's individual pursuits from the university. Having fostered their individual 'genius' in whatever avenue in science or art that they chose, they were now prepared to embark on a trip, each student visiting a specific geographical region, encountering what they were prepared for during their studies.[73]

Kelsall went on to describe what kind of studies each of his seven students had engaged in at the University, going as far as to give detailed accounts of their libraries, their final projects and descriptions of the places where each student would travel. However implausible his scheme may have seemed, some specific historical conditions can be identified which helped shape his fantasy. As he acknowledged, 'It is by a judicious division of labour [such as was just outlined] that the work of M. Humboldt will probably prove so interesting.' This was a reference to Wilhelm von Humboldt, who in 1809 had proposed suggestions for the reorganisation of university studies in Germany, in what was to be part of the framework for the foundation of the university of Berlin.[74] Humboldt was concerned about the maintenance of the administration of the 'moral culture of the nation'. In his view, the best structure for intellectual institutions would permit 'the combination of objective scientific and scholarly knowledge with the development of the person'. One repeated emphasis which Kelsall seemed most sympathetic to was 'freedom in [the students'] intellectual activities' and collaboration between each person's successful intellectual achievements. The best design of the university which would most benefit the students, and ultimately the state, was based upon 'the principle of cultivating science and scholarship for [the student's] own sake. . . . This is the secret of good research method.'[75]

Humboldt's emphasis on 'freedom of intellectual activities' in the new university was a highly political prescription (part of German Idealism that developed in response to Napoleon's invasion of Jena in 1806), the full extent of which Kelsall was not attempting to import, and which need not concern us here.[76] But the association Kelsall made between his ideal university and Humboldt's is interesting, not least as it points to some of the ways that ideas about curriculum reform, the division of intellectual labour and the commitments to a 'universe of knowledge' that would embrace sciences, history, art and classicism were shared by different commentators around Europe (a point we will return to below). Humboldt's ideas were directed to an institutional culture that later materialised. Kelsall's did not, but he did take up some of his own ideas. He also toured

Greece (and later other parts of Europe), collecting information which subsequently was given to Clarke, which he used to make some corrections to the second edition of his *Travels*. Kelsall was clearly taken with the effectiveness of Clarke's innovative teaching and travelling experience, a point further supported by Kelsall's dedication of his *Phantasm* to Clarke.[77]

Significant changes in pedagogy were made in the early nineteenth century, and the changes in forms and extent of foreign travel were not unrelated. Foreign trips seemed to be increasingly part of growing up. Now aristocrats travelled not for pleasure, but for diplomatic education. So Georgiana, Lady Spencer, would write to her daughter, Lady Bessborough, to soothe the latter's anxieties about her seventeen-year-old son's appointment as secretarial aide to the British Ambassador in Petersburg, Lord Granville Leveson-Gower:

> I think as you state it, my dear Harriet, there can be no objection to Willm's going to Russia. How far Ld. Gr. L. G. is qualified for diplomatic business you are a much better judge than I can be, but I conclude he has some secretary or experienced person goes [*sic*] with him from whom Willm may get information, & he should take this opportunity of acquainting himself with Modern History, especially that part of it that regards Prussia, Russia, Denmark, Sweden & Poland.[78]

Such trends continued and attracted wider appeal. As one reviewer of travel literature observed in 1806, the 'extreme partiality of our countrymen for travelling is a subject which has often excited the surprise of foreigners'.

> It would not be easy, perhaps, to explain the causes of this propensity; but there can be little doubt of the fact, that it exists among the English in a greater degree than among any other people. At the close of every term, our universities send forth their raw productions to be exported in abundance to the Continent; and no sooner is the season of fashionable gayety concluded in London, than the roads are covered with *tourists* and *travellers*, who issue from the metropolis in every direction.[79]

The swelling stock of information collected by eighteenth-century travellers which was presented to the nineteenth century generated more interest and prompted further comparative studies and more extensive, wideranging travel. Whether seeking simply to attract public attention to their own peregrinations, or provide more detail for the contemplation of philosophers, historians or politicians, by the early nineteenth century travellers needed to go further, take longer, or write and collect more. One traveller distinguished himself by claiming to have *walked* over much of the 'inhabitable globe'![80] But besides setting a new pace for future explorers, the achievements of the eighteenth-century travellers such as those discussed in this book also provided a number of ways that fellow British could think about European identities.

Looking at Europe from different perspectives

Throughout the eighteenth century, growing imperial concerns brought on a surge of interest in continental affairs, as travellers busily mapped the civilising process and assessed the limits of modernity abroad. By the century's end, central European states began to recognise in each other shared cultural values which differentiated them – as part of a similar intellectual, if not political, *community* – from non-European peoples as described by various travellers. One historian has recently summed up the development of a 'European self-confidence' during the Revolutionary–Napoleonic period:

> These years witnessed the construction of a cultural and political concept of Europe which was structured fundamentally around two perspectives. First, a European view of the extra-European world was consolidated which drew on earlier perceptions, but transformed them into a radically different unifying concept of European civilization and progress which allowed the classification, and justified the material exploitation, of the rest of the world. Secondly, a distinctive conviction was forged of what constituted the essence of Europe's superiority, based on a division of its land mass into nation states and the role of the rational state in furthering progress.[81]

The perception of the 'rest of the world' from the point of view of a unified 'European civilization' here refers mainly to European imperial explorations of Asia, India, and Africa.

The present study has sought to examine what might be considered the 'other side' of such common assumptions of 'European perceptions'. Following Edward Said, the historian quoted above continues to write that by the end of the eighteenth century, 'the European republic of letters had developed sophisticated tools with which to classify and understand extra-European societies'. The use of such analytical tools rendered intelligible 'for a European public the description of places and peoples which were *not only distant*, but often bordered on the fantastic'.[82] I would replace 'not only distant' with 'not *always* distant' in order to point out how close to home and how ambiguous the perceptions of European frontiers were. In addition, by considering ways that Europeans looked at Europe, we can begin to see how the use of such analytical tools not only *did* allow for the possibility of a *shared* European view of the 'non-European', but how classificatory projects could also work to support national claims to superiority. In other words, British travellers' reports about the state of Europe at times fed into nationalistic claims that Britain was more modern, civilised and enlightened than her imperial rivals.

Eighteenth-century travellers and the readers and critics of their published narratives researched and reflected upon questions similar to those presented at the start of this book. Where exactly was the 'outside' of Europe? What constituted an 'extra-European' world? In what ways could the Enlightenment 'sciences of man' be used to classify populations? Did travellers' inspections of foreign frontiers place more emphasis on natural or cultural conditions of life? To what was difference between populations attributed, and how far did similarities between diverse groups of people extend? Travellers' comparative studies of different peoples at the European frontiers helped provide answers to these questions, summaries of which would be useful for a conclusion.

While this book has been organised geographically – tracing the ways that travellers' discussed the characteristics of the northern, eastern and southern frontiers of Europe – it is also helpful to consider other ways that travellers and commentators conceived of the similarities and differences of diverse European populations. Among travellers' concerns were geographical and topographical diversity, climatic variations, and natural historical distinctiveness of different areas around Europe. These were among the *natural* differences that seemed to affect conditions of life in harsh, hot, frozen or barren places. They were the variable 'circumstances' or 'situations' beyond which one had to look in attempt to discern common 'natural characteristics' of the human species, 'constants' of human nature.[83]

Much attention was also given to cultural variations amongst the peoples encountered. These included different diets, dress, political and educational commitments, artistic patronage and social structures. Broadly speaking, we can observe that in attempts to map civilisation – in trying to distinguish degrees of Europeanness – the further from the 'centre' the more attention was paid to the different *natural* conditions of life. Here, attention to natural resources provided the data for understanding the material underpinnings of more 'primitive,' topographically distinct, 'non-European' populations. As the travellers looked 'inward,' towards metropolitan locations, the categories used to distinguish Europeanness relied more on assessments of *cultural* apparatus – from architecture to educational ideals. Thus, the question of what defined European status gradually became an assessment of what defined a certain *kind* of European. In particular, the evaluation of a European status was a matter of how the British assessed others' cultural achievements as part of a measure of what defined *modern* Europe. This ultimately contributed to a nationalist sentiment suggesting that Britain was *the* measure of all other degrees of modernity and civility.

British travellers assessed the three broad geographical areas that formed the structure for this book. They can be summarised in the following way. Scandinavia, principally appraised relative to the northern governments' economic stability, population health and commitment to the promotion of the arts and sciences, was illustrative of a once enlightened land suddenly in cultural decline. As was concluded at the end of that chapter, the once promising pursuit of an 'alternative enlightenment' – epitomised by the early success of its natural philosophers and practical educational programmes – failed in the decades following the bloodless revolution of 1772. Russia, despite a century of spirited attempts from Peter to Paul, was still only half-way up the civilising scale. 'Enlightenment' was never achieved in Russia, it was continually asserted. Worse still, the barbaric tendencies of the tsar who welcomed the nineteenth century looked, to some, to be putting Russia at risk of degenerating into as primitive a condition as its provinces.

Both of these areas, it was thought by British commentators, could *partly* be classified European – essentially due to their economic and military interactions with Britain. So could Greece. But here, a different rationale guided British perceptions of what constituted European civility. In the ancient lands, there was an historic association between the civilised status of modern Europe and the legacy of the country in which the principles of modern government were first pronounced. There were also conflicting interests in this land, which formed the frontier for competing European imperial powers. Hence, British explorers of the ancient lands analysed historical civilisation, not to prove it inherently European, but to lay privileged claim to it by personally identifying themselves – as *British* – with what they hoped to appropriate as their heritage. The British were sharpening their analytical tools used to explore the ancient lands in order to fight off competing claims to the right to possess and identify with the origin of European civilisation.

But, there are other ways to consider how British travellers conceived of European frontiers. In the brief account that follows, rather than reapproaching the three separate regions, we will look at areas around Europe as conceptual concentric spheres which embraced the frontier lands and gradually homed in on a British centre, to see how travellers theoretically classified European and non-European populations. The outermost 'sphere' represents the ambiguous frontier – populated with primitive Laplanders and Tartars: 'human kinds' who were most readily classified according to natural characteristics such as physiognomy, climate, language and migratory lifestyle.[84] As the spheres enclose areas increasingly closer to home, however, new values were deployed to evaluate the degree of civility ascribed to those under inspection. Such judgements fell within the realm of political, cultural and ideological critiques. Thus, looking at how the frontiers were conceptualised reveals varying ways that populations were classified from primitive to culturally refined; how they were distinguished in degrees from barbarity to modernity.

The history and diversity of mankind was a subject of inquiry for a broad range of Enlightenment theorists across Europe, led foremost by writers such as the Comte de Buffon and Anne Robert Jacques Turgot in France; the Swiss physiognomist Johann Caspar Lavater; William Falconer, David Hume, James Burnet (Lord Monboddo), Adam Ferguson and John Millar in Britain; Samuel Stanhope Smith in America; Carlus Linnaeus in Sweden; Samuel Pufendorf and Johann Friedrich Blumenbach in Germany; and the Anglo-German naturalist Johann Reinhold Forster – to list only a few.[85] Travellers' writings were systematically combed for material to provide the detail in the larger maps of humanity. In an age when the propagation of travellers' narratives was partly driven by a demand for accounts of the 'weird and wonderful', the Anglo-Irish traveller, orientalist and numismatist William Marsden, for one, felt it necessary to advise philosophers to select carefully from only the most reliable reports:

> Facts [must] serve as *data* in [philosophers'] reasonings, which are too often rendered nugatory, and not seldom ridiculous, by assuming the truths, the misconceptions, or wilful impositions of travellers. The study of our own species is doubtless the most interesting and important that can claim the attention of mankind; and this science, like all others, it is impossible to improve by abstract speculation, merely. A regular series of authenticated facts is what alone can enable us to rise towards a perfect knowledge in it.[86]

It was a warning of which many eighteenth-century travellers from Britain as well as other parts of Europe also took heed, and increasingly refined their observations to more detailed, descriptive and demonstrable forms of evidence. They thus increasingly contributed to debates regarding the origin, progress and classification of modern civilisation. To this end, they shared not only information, but ways of discussing and representing extra-European populations. Certain widely cited travelogues acted as models of how to travel, record information and construct images of life at the frontiers. The Danish traveller Carsten Niebuhr's *Travels through Arabia* (first published in 1772 as *Beschreibung von Arabien*; English edition 1792) set one influential example of how a team of travellers should employ enlightened principles of disciplined observation and systematically gather statistics, specimens, temperature readings, and so on. The more particular the observations the better. Thus the Swedish botanist, Anders Sparrman, wrote in his *Voyage to the Cape of Good Hope* (Swedish, 1783; English, 1785) that 'every authentic and well-written book of voyages and travels is, in fact, a treatise of experimental philosophy.'[87] A number of travellers wrote with similar conviction, such as Alexander Russell, surgeon to the Levant Company, who included a comprehensive analysis of the history and distribution of disease in the East in his well-known and widely referred to *Natural History of Aleppo* (1756). A common analytical thread that was woven through these accounts of European frontiers was a philosophical inquiry into human diversity. The work of antiquarians, philologists, political and religious historians, medical men and others was being synthesised into new forms of inquiry now commonly associated with leading Enlightenment philosophers. As the Scottish philosopher Henry Home (Lord Kames) commented in the 1770s, 'natural history, that of man especially, is of late years much ripened.'[88]

When travellers wrote about their journeys to the European frontiers, they at once contributed to a Eurocentric philosophy about the scientific and historical classification of human kinds, and used that philosophy to guide their descriptions. The images of the frontier that travellers constructed and the theories that made sense of life there were thus interdependent. At times it appeared shocking that more seemed to be known of geographically distant peoples – those in the South Seas, the New World or the Far East – than those who lived in relative proximity to Europe. The frontier land was a blank on the map of civilisation. No one from the central states of Europe had accurately charted the land; no study of language had illuminated cultural relations between various populations; no one had even reported on the practicalities of life in the varied climatic conditions. As late as the end of the eighteenth century, there was little to distinguish between the cultures of the Lapps, Tartars and Cossacks – they were grouped together into the general classification of 'primitive', nomads who lived in the ambiguous peripheral circle around Europe.

Different populations around the world were already generalised into groups and ranked on a broad scale according to the stage in the civilising process they had reached. William Marsden made distinctions between different ranks of civil society and classified different populations according to five classes: 'republics of ancient Greece, in the days of their splendor . . . France, England, and other refined nations of Europe'; 'the great Asiatic empires'; select 'states of the eastern archipelago'; 'less civilized Sumatrans [and] newly discovered people [in the] South Sea'; and finally the 'Caribs, New Hollanders, the Laplanders, and the Hottentots'.[89] In other systems the Laplanders were grouped with others at the European frontiers, more often than not being spoken about along side the Tartars, who were associated with the Asian Mongols and the Turks. When Clarke toured Lapland, he noted a number of similarities between the Lapps and the Tartars, from their appearance down to the custom of carrying their babies in strikingly similar cradles. He was not alone in speculating on a shared 'oriental' ancestry between the two groups – the philological studies already mentioned by William Jones or John Richardson being another example of new ways that these cultures were being analysed, categorised, and associated with a common point of origin.

What many commentators saw as a common denominator between these different groups that helped to classify them together was their struggle to survive in harsh climates. The environment, whether in Lapland, Tartar or Cossack territory, was 'rigorous'. Some theorists, following lines of reasoning similar to Richard Phillips' (discussed above), believed that it was the extreme environmental conditions that caused people to look as they did: the environment stimulated physiognomic 'types', further supporting classificatory groupings of people at the frontiers. An example of such reasoning is found in the 1787 tract by the American Presbyterian minister and professor at Princeton University, Samuel Stanhope Smith. In his *Essay on the Causes of the Variety of Complexion and Figure in the Human Species*, he argued that physical variety among people in different geographical areas was due to natural causes. He gave as an example the case of the Tartars and the Laplanders:

> The whole Tartar race, except a few small tribes who have probably migrated into that country from other regions, are of a lower stature than their southern neighbours on the continent of Asia, or than the people of the temperate latitutdes of Europe. Their heads are large; their shoulders raised; and their necks short; their eyes are small, and appear, by the great projection of the eye-brows, to be sunk in the head; the nose is short, and is not so prominent as the same feature in the Europeans; the cheek is elevated; the face, somewhat depressed in the middle, and spread out toward the sides; and the whole appearance and expression of the countenance is harsh and uncouth. All these deformities are aggravated as we proceed towards the pole, in the Laponian, Borandian, and Samoiede races, which, as Buffon justly remarks, are only Tartars reduced to the last degree of degeneracy. A race of men resembling the Laplanders in many of their lineaments and qualities, is found in a similar climate in America.[90]

And so through further testimony and travellers' observations came more data for the classification of types of populations around the world, where the same principles of environmental effect on physiognomy would apply. Because of the 'rigorous climates' in northern Tartary and Lapland, the inhabitants' noses were short since they drew breath through their noses, whereby the cold air numbed facial muscles, which tended 'to restrain the freedom of [that feature's] expansion'. Their foreheads were prominent features since the 'superior warmth and impulse of the blood in the brain, which fills the upper part of the head, will naturally increase its relative magnitude'. Their eyes appeared small due to the 'contraction of their lids occasioned by extreme cold'. And so on. Europeans, by contrast – perfectly represented in the archetype of Greek physique – who lived in 'the temperate zone', with the most agreeable warmth allowing for free and easy muscle expansion; their features had 'the most pleasing and regular proportions'.[91]

It should be noted that Smith was writing in opposition to a different theory, such as maintained by Lord Kames. In his *Sketches on the History of Man* (1778), Kames argued that climate could *not* account for physiognomic differences between humans. If different races were generated by climatic conditions, then how, he asked, could one account for the differences of appearance amongst different populations who live so closely together in northern Scandinavia? 'Lapland is piercingly cold, but so is Finland, and the northern parts of Norway, the inhabitants of which are tall, comely, and well proportioned,' so he had read.[92]

The further subtleties of the debates over the possible effects of climate on physiognomy are irrelevant to us. It was mentioned here to point out that in either philosophical system, the data accumulated through travel and foreign encounters were used to construct classification systems of types, or 'kinds' or 'races', of non-European peoples. But at the same time, the accounts and classification systems were used to create, reflexively, the classification of the European – not only in terms of similar physiognomy (in distinction to the features of those at the frontiers), but in terms of similarities between European locations (whether that being responsible for 'creating' the European or the European was divinely created for that location).

The debate over what role the environment played in shaping the physical and social identity of those at the frontier was further relevant to understanding the historical identity of Europeans. In Smith's model, the correlation between environmental conditions and the 'state of society' had significant implications regarding theories of the progress of civilisation. In his scheme, those classified by Europeans as 'savage' or 'primitive' were not inherently so, but were equally capable of becoming civilised peoples (as defined by European standards), through physical adaptation to different, more 'favourable', environments. In other words, it was conceivable that those at the European frontiers provided glimpses of modern European ancestry.[93] The eighteenth-century equivalent to L.P. Hartley's *bon mot* that 'the past is a foreign country' might have read: 'a foreign country is our past'.

Perhaps never more so than at the immediate European frontiers were travellers' observations so relevant to reflections about the constitution of a European identity. Reaching agreement about how best to classify those outside of Europe provided further criteria for the delineation of a European ('civilised' and 'symmetrical') 'environment'. The shared uses of cultural categories and terms of reference facilitated thinking of oneself as European, in the context of the classification of types of populations. But the closer those populations were to Britain, France or Germany, for example, where those who invented the languages of population classification lived, the more we find that the concept of a common European identity was difficult to sustain. Classifying Europeanness turned from the seemingly obvious grouping of those who were in a state of civilisation, to groupings of how people of different nations *act* – who was best suited to guide the *civilising process*.

As Lucien Febvre has shown, the evolving uses of the term 'civilisation' in French and British discourse gradually created the conception of various and competing 'European civilisation*s*', nationally oriented. In France in the 1760s, to speak of civilisation was to speak of a continuing process of refinement – of developing social laws and government, in opposition to 'barbarity'. So in Nicolas-Antoine Boulanger's posthumous *Antiquité dévoilée par ses usages* (1766), we have the statement: 'When a savage people has become civilised, we must not put an end to the act of *civilisation* by giving it rigid and irrevocable laws; we must make it look upon the legislation given to it as a form of *continuous civilisation*.'[94]

By the end of the century, the concept of guiding a 'savage people' to a civilised condition was laden with different theories of what ultimately constituted 'civilisation'. Such a status came to refer not only to the maintenance of social order (through policing and government), but to sustained wealth in philosophic, scientific, artistic and literary culture. To speak of the act of civilising – to civilise a population – became attached to the concept of moral, social and cultural progress. Thus, 'the European' was the civiliser – the creator of the state civilisation. Refining the classification of European civilisation also refined the category of 'the European'.

So, European visions of extra-European society were also laden with different nationalist expressions of what best 'to do' about those populations.

In the Europe of post-Revolutionary France and early industrial Britain, reflections on how best to think of European identity were imbued with notions of who best represented *the civilised*. As the last chapter pointed out, this was the underlying issue at stake in the contest between Britain and France over the imperial and historical frontier in the Levant. Control over that land and exclusive rights to its 'historical memoirs' was tied to debates over the proper constitution of a 'free government', the benefits of artistic patronage of the arts, and Britain or France's respective mastery over 'orientalism'. These concerns cast light on the emerging claims to identify a British – as opposed to a French, or broader European – civilisation.[95]

The same concerns applied to reflections about the degree of Europeanness of the Scandinavian states or Russia. Hence William Coxe's published views on the spread of enlightenment in Russia and the attention to the rate of their ongoing civilising process: 'their progress toward civilisation is very inconsiderable'; 'it is impossible even for a monarch ... to diffuse a love for the works of art among a people who must first imbibe a degree of taste'; 'the cultivation of a numerous and widely dispersed people is not the work of a moment, and can only be effected by a gradual and almost insensible progress'; and so on.[96]

The shared vocabulary that gave birth to 'civilisation' also invented Eurocentrism. But at the same time, debates over who best embodied and applied the principles of enlightenment and civil duty to social improvement further refined the categories of the European to a specific national level. Eurocentrism turned into 'enlightened nationalism'. The discourses of a 'European' or a 'British' identity were not self-evident. Particular kinds of intellectual work were required to create these reflexive references. To be 'British' relied, in this context, on the self-nominated and self-justified qualities that distinguished the interests of Britain above and beyond other European states. Similar claims were made from other nations. Thus the 'limits of Europe' were not merely geographically defined, but relied on distinctions between culturally created categories tracing varying degrees of social refinement, enlightened rule and civility.[97] The boundaries, like classification systems, were pliant. As John Richardson observed of the problems involved with general attempts to classify populations:

Men totally dissimilar are grouped together, under one indiscriminate character, merely because they are known in Europe
by one general name; whilst, among their numerous nations, a
difference of character may prevail, not inferior perhaps to that
which marks an *Englishman* from a *Frenchman*, a *Hollander* from
a *Portuguese*.[98]

The problems involved with classifying populations and of assigning boundaries (whether geographical, historical or biological) have
never gone away. But by exploring and exposing the diversity of
life at the frontiers, and pointing out the inherent insecurities in
mapping civilisation, eighteenth-century travellers took the first steps
in a two hundred-year journey destined to reconstruct European
identity.

Notes and References

Chapter 1

1 Critiques on the idea of Europe relative to the politics of identity include Delanty, *Inventing Europe*; Couloubaritsis et al., *The Origins of European Identity*; Hargreaves, 'European Identity and the Colonial Frontier'; more recent accounts which emulate the traditional travel narrative which seek to provide a geopolitical adventure to the heart of today's European Commission include Middleton, *Travels as a Brussels Scout*; Fraser, *Continental Drifts*.

2 Emerson, *Conduct of Life*.

3 Fox, Porter and Wokler, eds, *Inventing Human Science*; Moravia, 'The Enlightenment and the Sciences of Man'.

4 'Oriental' derives from the Latin verb *orior*, meaning 'to arise', its participle *oriens* used to denote 'the rising sun' or 'the east'. For its distinction with the *occident*, or place where sun sets as discussed relative to eighteenth-century travel literature, see Lowe, *Critical Terrains*, chapter 2.

5 See Wokler, 'Anthropology and Conjectural History', in Fox, Porter and Wokler, eds, *Inventing Human Science*.

6 Elias, *The Civilizing Process*.

7 Thomas, *Entangled Objects*.

8 Colley, *Britons*; Brewer, *Pleasures of the Imagination*; one work which attempts to fill the desideratum is Wolff, *Inventing Eastern Europe*, which looks at travellers from around Europe who wrote about Poland and Russia; for Asia see the superb study by Marshall and Williams, *The Great Map of Mankind*. For a study that deals with French and German travellers' perceptions of Europe, see Lowe, *Critical Terrains*.

9 Williams, *The Great South Sea*. Recent anniversaries of Cook and Banks have resulted in a variety of scholarship illuminating their life and work. For example, on Cook: Fisher and Johnston, eds, *Captain James Cook and His Times*; M. Hoare, ed., special issue of *Pacific Studies* 1 (1978), 71–212; on Banks: Banks et al., eds, *Sir Joseph Banks*; Gascoigne, *Joseph Banks and the English Enlightenment*. For more general accounts of voyages of exploration, see Miller and Reill, eds, *Visions of Empire*; Friis, ed., *The Pacific Basin*; Beaglehole, 'Eighteenth-Century Science'.

10 Pagden, *European Encounters with the New World*, particularly chapter 1.

11 Thomas, *Man and the Natural World*; Ritvo, *The Platypus and the Mermaid*.

12 Clarke, *Travels*, Vol. 9, pp. 468, 560; all references to Clarke's *Travels* refer to the 4th edition unless otherwise stated. Bell, *Travels*, Vol. 1, p. 30. Gibbon, *Decline and Fall of the Roman Empire*, Vol. 6, p. 450.

13 For discussion of anthropology and human classification schemes, see Marshall and Williams, *The Great Map of Mankind*; for its later manifestations, see Stepan, *The Idea of Race in Science*.

14 Ferguson, *Principles of Moral and Political Science*, Vol. 1, p. 97.

15 A glance at the popularity of voluminous 'collections of voyages' suggests the extent and range of travel literature; for example, Astley, *A New General Collection of Voyages and Travels*; Campbell, ed., *Navigantium atque Itinerantium Bibliotheca*; Pinkerton, *A General Collection of the best and most interesting Voyages*; Walpole, *Memoirs relating to European and Asiatic Turkey*. For British reading culture, see Brewer, *Pleasures of the Imagination*, chapter 4, especially pp. 180–1 for figures from the Bristol borrowing library.

16 Quoted in Collins, *Profession of Letters*, p. 55.

17 See Pimlott, *The Englishman's Holiday*, p. 68 for first; Cunningham, *Cautions to Continental Travellers*, p. 3; for recent figures, see Towner, 'The Grand Tour'; for general overview see Black, *The British Abroad*; Hibbert, *The Grand Tour*; for English Grand Tourists in France, see Cohen, 'The Grand Tour'. Colley, *Britons*.

18 Boswell, *The Life of Samuel Johnson*, p. 61.

19 Leask, *British Romantic Writers*; Hudson, *Writing and European Thought*; Butler, 'Romanticism in England,' in Porter and Teich, eds, *Romanticism in National Context*.

20 See Habermas, *The Structural Transformation of the Public Sphere*, for bourgeois constructs in literary and material culture. For discussion of the professionalisation of travel writing and contemporary concerns over truth telling and embellishment of travel accounts, see Pratt, *Imperial Eyes*, 86–8; Adams, *Travellers and Travel Liars*.

21 Gascoigne, *Cambridge in the Age of the Enlightenment*; Winstanely, *The University of Cambridge*; Searby, *History of the University of Cambridge*; Stone, *Universities*; Sutherland and Mitchell, eds, *History of the University of Oxford*.

22 For an account of a late eighteenth-century Cambridge traveller to the Pacific who falls within these categories, see Dening, *The Death of William Gooch*.

23 Clarke, *A Tour through the South of England*.

24 Otter, *Remains*, Vol. 1, p. 123.

25 Clarke, *Travels*, Vol. 1, p. 1.

26 Pratt, *Imperial Eyes*, for the popular genre of 'survival literature'.

27 Besterman, *The Publishing Firm of Cadell and Davies*.

28 Robinson, *Wayward Women*, pp. 194–5; Lister, *A Bibliography of Murray's Handbooks*; see also Vaughan, *The English Guide Book* and Buzard, *The Beaten Track*, for accounts of the rise of tourism.

29 Porter and Teich, eds, *The Enlightenment in National Context*; Porter and Teich, eds, *Romanticism in National Context*.

Chapter 2

1 Clarke, *Travels*, Volume 9, pp. 1–10; all references to Clarke's *Travels* are to the 4th edition, unless otherwise noted. James, *The Travel Diaries of Thomas Robert Malthus*, p. 27. References to Clarke's correspondence is from Otter, *Remains*: Clarke to Anne Clarke, Hamburg, 28 May 1799, in Otter, *Remains*, Vol. 1, p. 452 for time of travel, Clarke to

Anne Clarke, Copenhagen, 7 June 1799, in Otter, *Remains*, Vol. 1, p. 453 for quotation.

2 See Kirby, *Northern Europe*.

3 Roberts, *The Age of Liberty*.

4 Metcalf, 'The First "Modern" Party System?'.

5 Coxe, *Travels*.

6 Hallendorff and Schück, *History of Sweden*, pp. 222–6 for a sketch of Gustav Adolf's reign.

7 Coxe, *Travels*, Vol. 2, p. 333.

8 Quoted in Wright, 'Defoe's writings on Sweden', p. 25.

9 Upton, 'The Swedish Nobility', in Scott, ed., *The European Nobilities*.

10 Sheridan, *A History of the Late Revolution in Sweden*, pp. 3, 6.

11 Coxe, *Travels*, Vol. 2, p. 396.

12 Barton, 'Late Gustavian Autocracy in Sweden', p. 280.

13 Lucas, 'Great Britain and the Union of Norway and Sweden'; Ryan, 'The Defence of British Trade'.

14 Carr, *A Northern Summer*, p. 1.

15 Walpole, ed., *Travels in Various Countries of the East*, pp. iii–iv.

16 Clarke, *Travels*, Vol. 9, p. 43.

17 Clarke, *Travels*, Vol. 9, p. 180.

18 Clarke, *Travels*, Vol. 9, pp. 195–6.

19 Clarke to Robert Tyrwhit, Tornea, Sweden, 9 July 1799, in Otter, *Remains*, Vol. 1, pp. 461–2.

20 Millward, *Scandinavian Lands*, chapters 3 and 12; O'Dell, *The Scandinavian World*.

21 Lindqvist, 'Labs in the Woods', pp. 301–6; Lindqvist, 'Natural Resources and Technology'.

22 Linnaeus, *Flora Lapponica*; Pontoppidan, *Natural History of Norway*; Engeström, *Guide du Voyages*.

23 Clarke, *Travels*, Vol. 9, p. 108.

24 Beer, *The Romance*.

25 Labbe, 'A Family Romance', p. 212.

26 Wollstonecraft, *Letters Written during a Short Residence in Sweden*.

27 Clarke to Otter, Tronheim, Sweden, 23 September 1799, in Otter, *Remains*, Vol. 1, p. 472.

28 Linnaeus, *Species Plantarum*, p. 1.

29 Frängsmyr, ed., *Linnaeus*.

30 Coxe, *Travels*, Vol. 2, p. 447.

31 For the historiographic debate on the late eighteenth-century 'origins of modern science', see Cunningham and Williams, 'De-centring the "Big Picture"'.

32 Clarke, *Travels*, Vol. 9, p. 212.

33 Kent, *War and Trade in Northern Seas*, p. 125.

34 Roberts, *The Age of Liberty*, p. 215.

35 Carr, *A Northern Summer*, p. 123.

36 Coxe, *Travels*, Vol. 2, p. 330.

37 Clarke, *Travels*, Vol. 11, p. 19.

38 These boxes were in fact opened fifty years later by Erik Gustaf Geijer, and are now deposited at Uppsala University Library under the name

'Gustavianska samlingen'. Unfortunately I have not been able to confirm the contents of that collection.

39 Clarke, *Travels*, Vol. 9, p. 220.
40 Lundgren, 'The New Chemistry in Sweden', for further discussion of Gahn; Lindqvist, *Technology on Trial*, chapter 5 for Board of Mines.
41 Porter, 'The Promotion of Mining'.
42 For an account of the training regime in chemistry and mineralogy in Sweden, see Dolan, 'Transferring Skill', in Dolan, ed., *Science Unbound*.
43 Thomson, *Travels through Sweden*.
44 Clarke, *Travels*, Vol. 9, p. 108.
45 Sörlin, 'Scientific Travel', in Frängsmyr, ed., *Science in Sweden*, 96–123; Koerner, 'Purposes of Linnaean Travel', in Miller and Reill, eds, *Visions of Empire*, 117–52.
46 Clarke, *Travels*, Vol. 9, p. 108.
47 Clarke, *Travels*, Vol. 10, p. 441.
48 Clarke, *Travels*, Vol. 11, p. 37.
49 Briggs, *The Age of Improvement*; McNeil, *Under the Banner of Science*; Berman, *Social Change and Scientific Organization*.
50 See Fox, Porter and Wokler, eds, *Inventing Human Science*, for discussions on the background of the theories which shaped the 'sciences of man'.
51 Gigerenzer, et al., *The Empire of Chance*.
52 Baker, *Condorcet*.
53 Malthus, *Essay* (1798), pp. 165–9. References to the first edition (1798) or second edition (1803) will be indicated.
54 Malthus, *Essay* (1798), p. 193; see Avery, *Progress, Poverty, and Population*, for renewed general account of the different ways these writers tackled population problems and addressed issues of progress and human happiness.
55 Malthus, *Essay* (1798), p. 395; for natural theology and Christian demography, see Hilton, *The Age of Atonement*, especially chapter 3; also Gigerenzer, et al., *The Empire of Chance*, chapter 1 for Süssmilch; Bonar, *Malthus*, pp. 34–5 for Paley.
56 Quoted in Bonar, *Malthus*, p. 49; see also his chapter on the critics, pp. 355–99.
57 Price, *Observations on Reversionary Payments*; Hacking, *The Taming of Chance*, pp. 49–51.
58 Malthus, *Essay* (1803), p. 58.
59 Malthus's Norwegian diaries are still extant and have been published as James, ed., [hereafter referred to as 'Malthus'], *Travel Diaries*; his other diaries, covering Sweden and Russia, were lost after he lent them to Clarke, who used them to help write his own *Travels*.
60 Malthus, *Travel Diaries*, p. 208; Malthus, *Essay* (1803), p. 148.
61 Pontoppidan, *Natural History of Norway*, p. 245, quoted in Malthus, *Travel Diaries*, p. 145.
62 Malthus, *Travel Diaries*, p. 175.
63 Malthus, *Essay* (1803), p. 134; Tribe, *Land, Labour, and Economic Discourse*, chapter 5.
64 Malthus, *Essay* (1803), p. 150.
65 Malthus, *Essay* (1803), p. 150.

66 For discussion of this, see Bonar's notes to the first *Essay*, p. xx.
67 Malthus, *Essay* (1803), p. 157.
68 Malthus, *Essay* (1803), p. 158.
69 Malthus, *Essay* (1803), p. 153.
70 Malthus, *Essay* (1803), p. 159.
71 Johannisson, 'Why Cure the Sick?', in Brändström and Tedebrand, eds, *Society, Health and Population*, 323–30.
72 Phrase used by Charles Rosen to describe statistics on population and health; for general discussion, see his *A History of Public Health*, pp. 148–52; for revised history of population health, including an account of eighteenth-century Sweden, see Dorothy Porter, *Health, Civilisation, and the State*.
73 Rashid, 'Malthus's *Essay on Population*', for the credibility of Malthus's 'facts'; Malthus's disclaimer is in his own preface to the second edition.
74 Malthus, *Essay* (1803), p. 162.
75 Johannisson, 'The People's Health', in Dorothy Porter, ed., *The History of Public Health*, p. 167.
76 Malthus, *Essay* (1803), pp. 164–5.
77 Malthus, *Essay* (1798), pp. 83–4, in particular for his critique of the English Poor Laws.
78 For his qualification on this attack for voluntary charities, see Digby, 'Malthus and the Reform of the English Poor Law', in Turner, *Malthus and his Time*, pp. 157–69.
79 Malthus, *Essay* (1803), 164; in 1826 Malthus added to this conclusion an observation that, since the introduction of vaccination to Sweden in 1804 and due to the progress of agriculture and industry, the health and population of the country had both considerably advanced.
80 Modern scholarship, however, tells a different story. We now know that Malthus manipulated his statistics in order to provide 'factual' evidence to support his theory. See: Drake, 'Malthus on Norway', pp. 175–96; Rashid, 'Malthus's *Essay on Population*'; Patricia James's notes to the second edition of Malthus's *Essay* points out citation errors and discusses alternatives to the idea that he blatantly distorted his facts, see Malthus, *Essay on Population* (James, ed.).
81 Johannisson, 'Naturvetenskap på reträtt', brief English summary 153–4, for a discussion of the decline of natural science in Sweden in the second half of the eighteenth century.
82 See Moyne, *Raising the Wind*, especially chapter 3.
83 Trusler, *Habitable World Described*, Vol. 1, p. 184. For a summary of earlier accounts of Lapland, see the bibliography provided by Berchtold, *Patriotic Travellers*, Vol. 2, pp. 173–4.
84 Defoe quoted in Moyne, *Raising the Wind*, p. 81.
85 *Winter*, 1744 edition, in Zippel, *Thomson's Seasons: Critical Edition*, Text E, 11.843–4; 881–4; Thomson was inspired to write about the Lapps in later editions of *Winter* having read an account of them in Maupertuis, *The Figure of the Earth* (1738); see Hamilton, *Travel and Science*, pp. 87–103.
86 [John Campbell], *The Polite Correspondence*, p. 164. The work was published anonymously, but the eighteenth-century short-title catalogue

names John Campbell (1708–75) as the author, presumably since he was a well-known writer of fictional biography with themes of travel and politics.

87 Smollett, *Travels through France and Italy*, Letter xli, see also McKillop, 'Local Attachment and Cosmopolitanism', in Hilles and Bloom, eds, *From Sensibility to Romanticism*, p. 200.
88 Iliffe, '"Aplatisseur du Monde et de Cassini"'.
89 Collinder, *The Lapps*, p. 20.
90 Consett, *A Tour through Sweden*, pp. 44, 90 for quotations.
91 Clarke, *Travels*, Vol. 9, pp. 231, 519.
92 Clarke, *Travels*, Vol. 9, pp. 265, 269.
93 Clarke, *Travels*, Vol. 9, p. 326.
94 See discussion of this custom in Collinder, *The Lapps*, pp. 57–9.
95 Clarke, *Travels*, Vol. 9, p. 328.
96 Malthus, *Travel Diaries*, p. 182.
97 Malthus, *Travel Diaries*, pp. 188–189.
98 Malthus, *Travel Diaries*, p. 195.
99 In the eighteenth century, language studies were separated into philology ('love of words', referring to the study of literature and interpretation) and orthology (a branch of grammar, particularly the study of the proper and standardised use of words and names). 'Linguistics' as a broader science of language was not developed until the nineteenth century.
100 Andersson, 'Runes, Magic, and Ideology'.
101 Clarke, *Travels*, Vol. 9, p. 229.
102 Jones quoted in Bernal, *Black Athena*, p. 229; see Bernal's chapter 5 for more general discussion of 'romantic linguistics' and debates about linguistic proofs of cultural ancestry. Similar studies into linguistic ancestry later flourished through the activities of the Sanskrit scholar Friedrich Max Müller, who, in the latter part of the nineteenth century, held the chair of comparative philology at Oxford.
103 Clarke, *Travels*, Vol. 9, p. 230.
104 Broberg, 'Homo sapiens', in Frängsmyr, ed., *Linneus*, 156–94; for primate–primitives distinction, see also Morgan, 'Between Primates and Primitives'.
105 For 'wild children', see Douthwaite, 'Rewriting the Savage'; see discussion of debates on 'human diversity' in Smith, *History of Human Sciences*, chapter 8; for 'Jemmy', see Desmond and Moore, *Darwin*, p. 137.
106 Clarke, *Travels*, Vol. 9, p. 229.
107 Tyson, *Orang-Outang*. Although, the term 'pygmy' was also used by others, such as Comte de Buffon, in reference to 'dwarfish people': 'Deceived by some optical illusion, the ancient historians gravely mention whole nations of pygmies as existing in remote quarters of the world. The more accurate observation of the moderns, however, convinces us that these accounts are entire fabulous. The existence, therefore, of a pygmy race of mankind, being founded in error or in fable, we can expect to find men of diminutive stature only by accident, among men of the ordinary size.' For further comments on accounts of 'pygmies', see Firestone, *The Coasts of Illusion*, chapter 12.
108 Clarke, *Travels*, Vol. 9, p. 495.

109 Prichard, *Researches into the Physical History of Man*, pp. 7–8; see also Greene, *The Death of Adam*, chapter 8; MacCormack, 'Medicine and Anthropology', in Bynum and Porter, eds, *Companion Encyclopaedia*, pp. 1436–48.
110 Clarke, *Travels*, Vol. 9, p. 343.
111 Clarke, *Travels*, Vol. 9, p. 476; for 'conjectural history' of progress of civilisation, especially as defined by Enlightenment thinkers such as Ferguson, Hume and Smith, see Hont and Ignatieff, eds, *Wealth and Virtue* and Pocock, *Virtue, Commerce and History*.
112 Clarke, *Travels*, Vol. 9, pp. 560–1.

Chapter 3

1 Bassin, 'Russia between Europe and Asia'.
2 Khodarkovsky, 'From Frontier to Empire', p. 116; this is part of a special issue of *Russian History* devoted to 'The Frontier in Russian History'.
3 Kohut, 'Ukraine', in Greengrass, ed., *Conquest and Coalescence*; Subtelny, *A History of Ukraine*.
4 Hosking, *Russia*, pp. 3–45.
5 McNeill, *Europe's Steppe Frontier*.
6 Williams, *The Rise, Progress, and Present State of the Northern Governments*, Vol. 2, p. 115.
7 Kirby, *Northern Europe in the Early Modern Period*.
8 Fedorov, 'Russia and Britain in the Eighteenth Century', pp. 137–44.
9 Tooke, *View of the Russian Empire*, Vol. 1, p. vi; for more on the 'British Factory', see Cross, 'Chaplains to the British Factory in St Petersburg'.
10 Swinton, *Travels into Norway*, pp. 376–7.
11 De Bruin, *Voyage to the Levant*; Weber, *The Present State of Russia*; Strahlenberg, *An Historico-Geographical Description of the North and Eastern Parts of Europe*; see Cross, 'British Knowledge of Russian Culture'. A number of Professor Cross's wide-ranging articles on the theme of British knowledge of Russian culture are reprinted in Cross, *Anglo-Russica*. See also, however, his more recent *By the Banks of the Neva*. This account of Anglo-Russian relations has special emphasis on British diplomats, merchants and others resident in Russia during the eighteenth century.
12 Müller, *Voyages from Asia to America for completing the discoveries of the north-west coast of America*, volume 3 of Müller's original work, translated by Thomas Jeffreys; Gmelin, *Flora Sibirica, sive historia plantarum Sibiriae*; Pallas, *Travels through the Southern Provinces of the Russian Empire*, translated by Francis Blagdon in 2 vols for his *Modern Discoveries*; see also summary by M.S. Anderson, *Britain's Discovery of Russia*, pp. 85–8; Wolff, *Inventing Eastern Europe*, chapter 4, 'Mapping Eastern Europe: Political Geography and Cultural Cartography'.
13 Black, *Maps and History*.
14 Clarke, *Travels*, Vol. 1, pp. ix–x.
15 Cross, 'The Reverend William Tooke's Contributions to English Knowledge of Russia'; Cross, *By the Banks of the Neva*, pp. 109–13.

16 Tooke, *View of the Russian Empire*, Vol. 1, p. 303.
17 Quoted from Hosking, *Russia*, p. 29.
18 Williams, *Rise, Progress, and Present State*, Vol. 2, p. 313.
19 Clarke, *Travels*, Vol. 1, pp. vii–viii.
20 Clarke, *Travels*, Vol. 1, p. 337.
21 Clarke, *Travels*, Vol. 1, pp. 337–8.
22 Goldsmith, *The Citizen of the World*, pp. 240–1.
23 Quintana, *Oliver Goldsmith*.
24 Ferguson, *An Essay on the History of Civil Society*, pp. 112–13.
25 Marshall, *Travels through Holland, Flanders, Russia*, Vol. 3, pp. 132–8; for suspicion that Marshall was an armchair traveller who never in fact saw Russia, see Wolff, *Inventing Eastern Europe*, p. 81, who cites the opinion of John Parkinson in the 1790s (see reference below).
26 Wraxall, *A Tour through some of the Northern parts of Europe*, pp. 228–9; Marshall and Wraxall are also discussed in this context by Anderson, *Britain's Discovery*, p. 139.
27 [Sinclair], *General Observations regarding the Present State of the Russian Empire*, p. 17; Cross, 'Sir John Sinclair's *General Observations*', in Cross, *Anglo-Russica*, 51–61.
28 *The Danger of the Political Balance of Europe, translated from the French of the King of Sweden, by the Rt. Hon. Lord Mountmorres*, frequently attributed to Gustav III, but Anderson points out it was probably penned by the Prussian ambassador in Stockholm, Anderson, *Britain's Discovery of Russia*, p. 154.
29 *Danger of the Political Balance*, p. 239.
30 Clarke, *Travels*, Vol. 2, pp. 173–4.
31 In fact, Catherine appears to have been sensitive to conduct in the Crimea, and ordered Potemkin to dissociate Russian involvement from the ruthless political regime of Shagin Girey, the ruling kahn; see De Madariaga, *Russia in the Age of Catherine the Great*, pp. 388–98, 633 n.40.
32 Clarke, *Travels*, Vol. 2, pp. 206–7.
33 Said, *Orientalism*; see also uses of Said's thesis in Wolff, *Inventing Eastern Europe*.
34 Russell, *The History of Modern Europe*, Vol. 5, p. 330.
35 Hughes, *Russia under Peter the Great*; Oliva, *Russia in the Era of Peter the Great*, for summary of his travels and reforms.
36 Hosking, *Russia*, chapter 2, for Peter's 'secular state'.
37 Rice, 'The Conflux of Influences in Eighteenth-century Russian Art and Architecture', in Garrard, ed., *The Eighteenth Century in Russia*, pp. 267–99.
38 Dmytryshyn, ed., *The Modernisation of Russia*.
39 Putnam, ed., *Seven Britons in Imperial Russia*, has biographical information and lengthy extracts from the travel narratives of John Perry, Jonas Hanway, William Richardson, Sir James Harris, William Coxe, Robert Ker Porter and Sir Robert Thomas Wilson.
40 Tooke, *View of the Russian Empire*, Vol. 1, p. xi.
41 Perry, *The State of Russia*, qtd. in Putnam, ed., *Seven Britons in Imperial Russia*, p. 61. Peter required all nobles who owned more than thirty serfs to move to the new capital, conscripting 350 nobles, in addition

to 300 merchants and craftsmen; see De Madariaga, 'The Russian Nobility in the Seventeenth and Eighteenth Centuries', in Scott, ed., *The European Nobilities*, Vol. 2, chapter 8, pp. 263–4.

42 Jones, 'Urban Planning', in Garrard, ed., *Eighteenth Century in Russia*, pp. 338–9.

43 Perry in Putnam, ed., *Seven Britons in Imperial Russia*, p. 38.

44 Perry in Putnam, ed., *Seven Britons in Imperial Russia*, pp. 39–40.

45 Hanway, *An Historical Account of British Trade*, qtd. in Putnam, ed., *Seven Britons in Imperial Russia*, pp. 88–9.

46 Cross, 'The British in Catherine's Russia', in Garrard, ed., *Eighteenth Century in Russia*, pp. 235–6; Parkinson's manuscript travel diaries have been edited and published for the first time by William Collier as *A Tour of Russia, Siberia, and the Crimea 1792–1794* as part of the 'Russia through European Eyes' series, general editor Anthony Cross. This series also includes reprints of John Perry (1716), Friedrich Christian Weber (1722–3), Patrick Gordon (mss diary, 1635–99), Baron von Haxthausen (1856), William Richardson (1784), J. Hamel (1854), P.H. Bruce (1782) and C.H. Pearson (1859).

47 Swinton, *Travels into Norway*, pp. 229, 230.

48 Richard, *A Tour from London to Petersburg*, p. 28; for the suspicion that Richard was an armchair traveller who never set foot in Russia, see Cross, *By the Banks of the Neva*, p. 385.

49 Qtd. in Anderson, *Britain's Discovery*, p. 81; for Bentham's residence in Russia, see Christie, *The Benthams in Russia*, pp. 37–9 for the 'English quarter'.

50 Qtd. in Anderson, *Britain's Discovery*, p. 82.

51 Swinton, *Travels into Norway*, p. 231.

52 Hosking, *Russia*, pp. 85–8 for building the new capital, and p. 89 for Peter's eager plans to found the Academy.

53 Swinton, *Travels into Norway*, p. 389.

54 Coxe, *Travels*, Vol. 2, p. 194; Anderson, *Britain's Discovery*, p. 88.

55 Raeff, 'The Enlightenment in Russia', in Garrard, *Eighteenth Century in Russia*, 25–47, for survey of Enlightenment cross-currents.

56 As characterised by Clarke, *Travels*, Vol. 1, p. 69.

57 De Madariaga, *Russia in the Age of Catherine the Great*, chapter 1.

58 Grey, *Catherine the Great*; Cronin, *Catherine*.

59 Tooke, *View of the Russian Empire*, Vol. 1, pp. xv–xvi.

60 Kingston-Mann, *In Search of the True West* (which documents Russian efforts to appropriate Western solutions to the problem of economic backwardness since the time of Catherine the Great); Cronin, *Catherine*, chapter 19 for broader discussion of the 'literary scene'.

61 Marker, *Publishing, Printing, and the Origins of Intellectual Life in Russia*.

62 Cronin, *Catherine*, p. 228.

63 De Madariaga, *Russia in the Age of Catherine the Great*, chapter 21 for discussion of 'court and culture'.

64 Coxe, *Travels*, Vol. 1, pp. 339–42; for an assessment of the diplomatic relations between Britain and Russia through a study of Harris see De Madariaga, *Britain, Russia, and the Armed Neutrality of 1780*.

65 Okenfuss, *The Rise and Fall of Latin Humanism*.

66   Cross, '"S anglinskago": Books of English Origin in Russian Translation';
      Cronin, *Catherine*, p. 233 for Sterne.
67   De Madariaga, *Russia in the Age of Catherine the Great*, p. 330.
68   Clarke, *Travels*, Vol. 1, pp. 92–3; 'Squire' Western refers to the charac-
      ter Squire Western in Fielding's *Tom Jones*.
69   Clarke, *Travels*, Vol. 1, p. 93.
70   Clarke, *Travels*, Vol. 1, pp. 86–7.
71   Swinton, *Travels into Norway*, pp. 456–7.
72   Carr, *A Northern Summer*, p. 354; see Wilson, *Muscovy*, pp. 158–62 for
      biographical remarks on Carr.
73   De Madariaga, *Russia in the Age of Catherine the Great*, chapter 10 for
      the 'Great Instruction', pp. 159–61 for discussion of debates over serfdom.
74   Williams, *Rise, Progress, and Present State*, Vol. 2, p. 304.
75   Richardson, *Anecdotes of the Russian Empire*, p. 371; see also informa-
      tion on Richardson in Putnam, ed., *Seven Britons*, chapter 3.
76   Quoted in Putnam, ed., *Seven Britons*, p. 164.
77   Carr, *A Northern Summer*, p. 355.
78   De Madariaga, *Russia in the Age of Catherine the Great*, pp. 542–3.
79   Alexander, *Emperor of the Cossacks*, pp. 56–60.
80   Carr, *A Northern Summer*, pp. 348–9.
81   For more detailed assessment of Paul's 'mixed emotions' as emperor,
      see Ragsdale, *Tsar Paul and the Question of Madness*.
82   Anderson, *Peter the Great*.
83   Saul, 'The Objectives of Paul's Italian Policy'.
84   My account is much too brief to reveal all the complexities of these
      political tensions. See: Schroeder, 'The Collapse of the Second Coali-
      tion'; McGrew, *Paul I*, pp. 309–21.
85   There is some debate whether Paul's actions were suddenly directed
      to seize upon British vulnerabilities in India; whether they were prompted
      by Napoleon to further the sanctions on trade; or a calculated plan
      for Russian geo-political expansion. See McGrew, *Paul I*, p. 316, for
      assessment of the evidence. For Paul's assassination, see Kennedy, 'The
      politics of assassination'.
86   Clarke, *Travels*, Vol. 1, pp. 6, 47.
87   Clarke, *Travels*, Vol. 1, pp. 130–1; for Clarke's critique of the account
      of Russia in the *Modern Universal History*, see p. 133; for these volumes
      as set readings in Cambridge when Clarke was writing his travels, see
      Smyth, 'A List of Books recommended and referred to in the Lectures
      on Modern History'.
88   Clarke, *Travels*, Vol. 1, pp. 7, 10–11.
89   Clarke, *Travels*, Vol. 1, p. ii.
90   Henry Brougham in *Edinburgh Review* 16 (1810), 334–68, p. 335.
91   *Anti-Jacobin Review* iii (1799), pp. 107–8; iv (1799), p. 476; v (1800),
      p. 112.
92   Clarke to Miss Newling, 18 October 1813, Cambridge University Library,
      MSS ADD 7082, f. 73; 'The most barbarian of Christians' is my trans-
      lation of the original 'INTER CHRISTIANOS ΒΑΡΒΑΡΩΤΑΤΟΙ.'
93   Letter from Brougham to John Allen, [31] March 1800, quoted in Jacyna,
      Philosophic Whigs, p. 36.

94 Malthus quoted in Bonar, *Malthus and his Work*, p. 415.
95 Maria Guthrie was the only one of the two to publish an account of Russia, *A Tour performed in the years 1795–1796*, quotation from Anderson, *Britain's Discovery*, pp. 96–7; see also Appleby, 'British Doctors in Russia', for a chapter on Matthew Guthrie.
96 See correspondence in British Library, MSS ADD 30,106.
97 Lyall, *The Character of the Russians*, pp. v–xxiii; 294, 456.
98 Williams, *The Rise, Progress, and Present State*, Vol. 2, p. 329.
99 Karamzin quoted in Hosking, *Russia*, p. 86. For other assessments of eighteenth-century Russia, see Bartlett and Hartley, eds, *Russia in the Age of Enlightenment*.
100 Clarke, *Travels*, Vol. 1, pp. 227–8.
101 Clarke, *Travels*, Vol. 1, pp. 295–6.
102 Clarke, *Travels*, Vol. 1, p. 298.
103 De Madariaga, *Russia in the Age of Catherine the Great*, chapter 4.
104 Clarke, *Travels*, Vol. 1, p. 299.
105 Tickell, 'On the prospect of Peace', lines 175–6.
106 Clarke, *Travels*, Vol. 1, pp. 301–2.
107 Clarke, *Travels*, Vol. 1, pp. 302–3.
108 Clarke, *Travels*, Vol. 1, p. 310.
109 Clarke, *Travels*, Vol. 1, p. 312.
110 Clarke, *Travels*, Vol. 1, p. 321.
111 Shaw, *General Zoology*.
112 Clarke, *Travels*, Vol. 1, p. 347.
113 Clarke, *Travels*, Vol. 1, p. 353.
114 Clarke, *Travels*, Vol. 1, p. 396.
115 Clarke, *Travels*, Vol. 1, pp. 368–79.
116 Gibbon, *The Decline and Fall of the Roman Empire*, Vol. 6, p. 450; Herder, *Outlines of a Philosophy of the History of Man*, both discussed in Wolff, *Inventing Eastern Europe*, pp. 295–315.
117 Hanway, *An Historical Account of British Trade*; Coxe, *Travels*, Vol. 2, pp. 338–48.
118 Fisher, *Crimean Tatars*; Rorlich, *The Volga Tatars*.
119 Clarke, *Travels*, Vol. 1, p. 399.
120 Clarke, *Travels*, Vol. 1, pp. 403–4.
121 Clarke, *Travels*, Vol. 1, p. 410.
122 Clarke, *Travels*, Vol. 2, p. 319.
123 For discussion of trade relations and 'wars of monopoly', see Bayly, *Imperial Meridian*, pp. 58–9 for quotations; for early British trade relations with the Muscovy Company in the Caspian region, see Lawson, *The East India Company*.
124 Bayly, *Imperial Meridian*, p. 154 for Montesquieu; Clarke, *Travels*, Vol. 2, pp. 323–4.
125 Ferguson, *Essay on the History of Civil Society*, pp. 75–6.
126 Ferguson, *Essay on the History of Civil Society*, p. 97.
127 Quoted in Treasure, *The Making of Modern Europe*, p. 572.

Chapter 4

1 Beaucour, Laissus, and Orgogozo, *The Discovery of Egypt*; Lefebvre, *The French Revolution from 1739 to 1799*, pp. 218–21; Connelly, *The French Revolution and Napoleonic Era*, pp. 192–8.
2 Tregaskis, *Beyond the Grand Tour*, p. 39.
3 See C. Emsley, *The Longman Companion to Napoleonic Europe*, for further chronology.
4 Clarke, *Travels*, Vol. III, p. 336.
5 For histories of the 'ancients versus moderns debate' from Renaissance humanism to the Enlightenment, see, for example, Burke, *The Renaissance Sense of the Past*; Jones, *Ancients and Moderns*; Levine, *The Battle of the Books*; Kors and Korshin, eds, *Anticipation of the Enlightenment in England, France, and Germany*.
6 Sheridan, *A Plan of Education for the Young Nobility and Gentry of Great Britain*, p. 7.
7 Pocock, *The Machiavellian Moment*; idem, 'Gibbon's *Decline and Fall* and the World View of the Late Enlightenment', in idem, *Virtue, Commerce, and History*, pp. 143–56.
8 Gibbon, 'General Observations on the Fall of the Roman Empire in the West', in idem, *The Decline and Fall of the Roman Empire*, Vol. IV.
9 Burrow, *A Liberal Descent*.
10 Volney, *Les Ruines*, translated as *The Ruins: or, A Survey of the Revolutions of Empire*.
11 Volney, *The Ruins*, p. 12.
12 Volney, *The Ruins*, pp. 84–5.
13 Rigby, 'Volney's Rationalist Apocalypse', in Barker, et al., eds, *1789: Reading, Writing, Revolution*, pp. 22–37.
14 Marshall, 'Empire and Authority in the Later Eighteenth Century'; Kriegel, 'Liberty and Whiggery in Early Nineteenth-century England'.
15 Lawson, *The East India Company*, p. 66.
16 Bayly, *Imperial Meridian*, chapter 4.
17 Hume, *Essays: Moral, Political, and Literary*, pp. 120–1.
18 Darnton, *The Business of the Enlightenment*; Lough, *The Encyclopédie*.
19 Quoted in Augustinos, *French Odysseys*, p. 23.
20 Constantine, *Early Greek Travellers*, p. 12.
21 Haskell and Penny, *Taste and the Antique*, p. 100.
22 Quoted in Constantine, *Early Greek Travellers*, p. 101; see also his chapter 5 for further discussion of 'Winckelmann and Greece'.
23 For more on Winckelmann, see Leppmann, *Winckelmann*; Hatfield, *Winckelmann and his German Critics*.
24 Turner, *The Greek Heritage in Victorian Britain*, esp. pp. 40–3.
25 Dallaway, *Anecdotes of the Arts in England*; quoted in Stoneman, *Land of Lost Gods*, p. 112.
26 Haynes, *The Arundel Marbles*; Rowse, *Four Caroline Portraits*, chapter 5.
27 Constantine, *Early Greek Travellers*, p. 30.
28 Clarke, *Critique on the Character and Writings of Sor George Wheler*, p. 2; Stoneman, *Land of Lost Gods*, p. 80.
29 Constantine, *Early Greek Travellers*, p. 33.

30 Cust and Coluin, *History of the Society of Dilettanti*, pp. 125–6.
31 Quoted in Stoneman, *Land of Lost Gods*, p. 122.
32 Quoted in Stoneman, *Land of Lost Gods*, p. 123.
33 Stuart and Revett, *Antiquities of Athens* (Vol. I, 1762; Vol. II, 1788 (though dated 1787); Vol. III (edited by Willey Reveley), 1794; Vol. IV, 1816; Supplement, 1830); Crook, *The Greek Revival*, pp. 15–20, for influence of Stuart and Revett; Brewer, *The Pleasures of the Imagination*, 256–60, for the Society of Dilettanti.
34 Chandler, *Travels in Asia Minor*; Chandler, *Travels in Greece*; these two works were reprinted in a single volume as *Travels in Asia Minor, and Greece; or, an Account of a Tour made at the Expense of the Society of Dilettanti*, 2 vols in one (London, 1817, and again in 1825). I use the 1817 edition. Quote from Vol. I, p. viii.
35 'Instructions', in Chandler, *Travels*, Vol. I, pp. ix–x.
36 See Constantine's detailed examination of Chandler's uses of his predecessor's works, *Early Greek Travellers*, chapter 9.
37 Chandler, *Travels*, Vol. II, p. 124.
38 Chandler, *Travels*, Vol. II, pp. 334–5.
39 Chandler, *Travels*, Vol. I, p. 121.
40 Chandler, *Travels*, Vol. I, p. 44; Constantine, *Early Greek Travellers*, p. 9.
41 Spate, *French Painting*, p. 7.
42 Quoted in Stoneman, *Land of Lost Gods*, pp. 130–1.
43 Chandler, *Travels*, Vol. I, p. 27; see also Vol. II, p. 189.
44 Gibbon, *The Decline and Fall of the Roman Empire*, Vol. VI (1898), p. 486.
45 Winckelmann quoted in Constantine, *Early Greek Travellers*, p. 226.
46 Hughes, *Travels in Greece and Albania*, quoted in Woodhouse, *The Philhellenes*, p. 27; see also Angelomatis-Tsougarakis, *The Eve of the Greek Revival*, chapter 3.
47 These comments characterise the views of the majority of mid-eighteenth-century British travellers, but not *all* travellers to Greece. As Constantine (*Early Greek Travellers*, chapter 7) points out, in the 1770s the French traveller Pierre Augustin Guys stressed the continuities visible between ancient and modern Greece; Constantine does add, however, that 'the most generally received view, especially among Europeans at home, was that the modern Greeks had fallen away utterly from the ancient and that all connection and continuity were lost' (p. 150).
48 Chandler, *Travels*, Vol. II, p. 137.
49 Cook, *The Elgin Marbles*, p. 53.
50 St Clair, *Lord Elgin and the Marbles* (3 editions: 1967, 1983, 1998; I refer to the 1983 edition), p. 8. St Clair provides what is perhaps the most comprehensive treatment of the politics of the marbles. See also, however, the lengthy and detailed article by Arthur Hamilton Smith, 'Lord Elgin and His Collection'. Also recently published are Hitchens, *The Elgin Marbles*, and Vrettos, *The Elgin Affair*, both of which rely heavily on the above works. One book on this subject that has not received the attention and scholarly credit that it deserves is Jacob Rothenberg's *'Descensus ad Terram': The Acquisition and Reception of the Elgin Marbles*.

51 St Clair, *Lord Elgin and the Marbles*, p. 89.
52 Quoted in Cook, *Elgin Marbles*, p. 55.
53 Augustinos, *French Odysseys*, pp. 161–3.
54 See Kennedy, *A Cultural History of the French Revolution*, chapter 8, 'Vandalism and Conservation' for discussions of the committees on art in the early republic.
55 Augustinos, *French Odysseys*, p. 18.
56 Leith, *The Idea of Art as Propaganda in France, 1750–1799*.
57 McClellan, *Inventing the Louvre*, pp. 91–2; see also Cantarel-Besson, *La naissance*, for more detailed discussion of the organisation of the Louvre in the early republic.
58 McClellan, *Inventing the Louvre*, p. 116.
59 Hunt, *Politics, Culture, Class*, p. 92.
60 Spate, *French Painting*, pp. 10, 11 for respective quotations.
61 McClellan, *Inventing the Louvre*, p. 94.
62 Ouzouf, *Festivals of the French Revolution*.
63 Quoted in McClellan, *Inventing the Louvre*, p. 121.
64 Lindsay, *Death of the Hero*, p. 112.
65 Clarke to William Otter, 14 September 1801, in Otter, *Remains*, Vol. II, p. 145.
66 Clarke to Otter, 15 December 1801, in Otter, *Remains*, Vol. II, p. 158.
67 Smith, 'Lord Elgin', p. 204.
68 Clarke, *Travels*, Vol. 6, pp. 223–4.
69 Hitchens, *Elgin Marbles*, p. 49; Vrettos, *Elgin Affair*, p. 62; Woodhouse, *The Philhellenes*, p. 13; Stoneman, *Land of Lost Gods*, pp. 168, 174.
70 Hamilton, *Memorandum* (1811; 1815); Clarke refers to the 1811 edition.
71 Clarke, *Travels*, Vol. 6, p. 232.
72 Clarke, *Travels*, Vol. 6, p. 616. Today the statue of Ceres is known as Kistophoros, and adorns a central location in the Greek section of the Fitzwilliam Museum, Cambridge, along with many other marbles donated to the University by Clarke and Cripps.
73 Chandler, *Travels*, Vol. II, pp. 201, 204.
74 See chapter 5.
75 Dallaway, *Constantinople Ancient and Modern*.
76 Quoted in Tregaskis, *Beyond the Grand Tour*, p. 40; see also Tweddell, *Remains of John Tweddell*.
77 Liscombe, *William Wilkins*, p. 106; also see Wilkins' publications listed in bibliography.
78 Dodwell, *Classical and Topographical Tour through Greece*, Vol. I, pp. iv, 78.
79 See Gell's publications listed in bibliography.
80 Crook, *Greek Revival*, p. 34; Stoneman, *Land of Lost Gods*, pp. 155–60.
81 Hans, *New Trends in Education in the Eighteenth Century*, p. 53; for mathematics at Cambridge, see Gascoigne, *Cambridge in the Age of Enlightenment*.
82 J.H. Monk and C. Blomfield, eds, *Museum Criticum; or, Cambridge Classical Researches*, 7 nos (Cambridge, 1814–26), No. 1, p. v, and No. 6, p. 128, respectively.
83 Watkin, *Thomas Hope* and Liscombe, *William Wilkins*, respectively.
84 Holland, *Travels*, pp. 414, 407.

85 Miller, *The English in Athens*, p. 25; Crook, *Greek Revival*, p. 42.
86 Gell, *Narrative of a Journey*, p. 164.
87 Holland, *Travels*, p. 274; Gell, *Narrative of a Journey*, p. 168.
88 Holland, *Travels*, p. 274.
89 Holland, *Travels*, p. 531.
90 Crook, *The Greek Revival*, p. 39.
91 Smith, 'Lord Elgin'.
92 Hitchens, *Elgin Marbles*, pp. 39–40.
93 Rothenberg, *Descensus as Terram*, p. 370.
94 Jenkins, *Archaeologists*, p. 82.
95 Quoted in *The Monthly Magazine*, 1 May 1816, p. 352. Rothenberg, in *Descensus ad Terram*, points out that the legislature's consideration of the judgements of an artistic tribunal was an unprecedented political event, and one which made judgements regarding British taste, artistic value, and the importance attached to classical scholarship a matter of public debate. Furthermore, it is significant that parliamentary discussions about the acquisitions of the Parthenon marbles, which had begun in 1815, were only postponed because of the Battle of Waterloo in June of that year! See Rothenberg, p. 382.
96 Quoted in Constantine, *Early Greek Travellers*, p. 133.
97 Winckelmann, *The History of Ancient Art*, preface.
98 Hobhouse, *A Journey through Albania*, Vol. I, pp. 345–8.
99 Marchand, *Byron's Letters*, Vol. II, p. 66.
100 Marchand, *Byron's Letters*, Vol. II, pp. 65–6.
101 See Leask, *British Romantic Writers*.
102 This is a widely reprinted illustration: see, for example, Caygill, *British Museum*, p. 21; the original is in the Department of Prints and Drawings, British Museum, DG 12787.
103 Crook, *Greek Revival*, p. 113; see also Leoussi, *Nationalism and Classicism*, for other, later nineteenth-century expressions of the revival of Hellenism.

Chapter 5

1 *Cambridge Chronicle and Journal*, 13 November 1802.
2 Otter, *Remains*, Vol. 2, p. 168.
3 De Beer, *Sir Hans Sloane*, p. 152; Miller, *That Noble Cabinet*; Caygill, *The Story of the British Museum*, p. 7, for list.
4 Clarke, *Travels*, Vol. 1, p. 173. All references to Clarke's *Travels* refer to the 4th edition, unless otherwise stated.
5 Russell, 'John Henry Heuland'.
6 Simmonds, 'The Founders,' pp. 228–31; Smith, 'First Hundred Years', pp. 241–2.
7 Smith, 'First Hundred Years', p. 243.
8 Caygill, *British Museum*, p. 17.
9 [Anon], *Description of . . . the residence of John Soane*; Watkin, *Sir John Soane*, p. 424.
10 Dick, *A Descriptive Catalogue of a Museum of Antiquities*.

11 Bullock, *A Companion to the Liverpool Museum*.

12 An account of Lever's collection was written by the Keeper of the Department of Natural History and Modern Curiosities at the British Museum, George Shaw: *Musei Leveriani explicatio*; see also Ella, *Visits to the Leverian Museum*.

13 Dyer, *The Privileges of the University of Cambridge*, Vol. 2, p. 34.

14 Harraden, *Cantabrigia Depicta*, p. 179; also reported in the *Cambridge Chronicle and Journal*, 11 December 1802, p. 2.

15 From Clarke, *Greek Marbles*, Preface.

16 See article on Edward Daniel Clarke in Turner, ed., *The Dictionary of Art*, Vol. 7, p. 378.

17 Brock and Curthoys, eds, *History of University of Oxford*, Vol. 6, pp. 588–9; Gaisford, *Catalogus*.

18 Clarke, *Testimonies*; Clarke, *The Tomb of Alexander*; Clarke, *Greek Marbles*.

19 See the Literary Correspondence of Thomas James Mathias, British Library, MSS ADD 22976, ff. 234, 238, 240, 242, 244, 252, 268, 275, 277; also the Aberdeen Papers, MSS ADD 43229, ff. 177b, 240, 254 for Aberdeen and f. 239 for Gell.

20 Brougham in a review of Heriot, *Travels in the Canadas*, in *Edinburgh Review* 12 (1808), 212–25, p. 223.

21 Henry Brougham, review of the first volume of Clarke's *Travels*, in *Edinburgh Review* **16** (August 1810), 334–68, p. 336.

22 Byron was at Trinity from 1805 to 1807. For his introduction to Clarke, see Quennell, *Byron*, Vol. 1, p. 118.

23 Hobhouse to Byron, 10 December 1810, printed in Graham, ed., *Byron's Bulldog*, p. 67.

24 Byron to Hobhouse, 22 October 1811, printed in Quennell, *Byron*, Vol. 1, p. 124.

25 Hobhouse to Byron, 15 July 1811, printed in Graham, ed., *Byron's Bulldog*, p. 93.

26 Byron to Hobhouse, 3 November 1811, printed in Marchand, ed., *Byron's Letters and Journals*, Vol. 1, p. 125.

27 Hobhouse, *A Journey through Albania*.

28 Young, 'Malthus and the evolutionists'.

29 The publishers Cadell and Davies, for example, churned out a lengthy list of books on history, politics and travel. They also produced a short-lived periodical titled *The Imperial Review*, meant to examine works relating to matters of the empire, which contained over 90 per cent 'reviews' of books published by their own firm.

30 From his *Advice to an Author* (1710), quoted in Frantz, *The English Traveller*, p. 8.

31 John Hawkins, 'Probationary Odes for the Laureateship'.

32 Oddy, *European Commerce*.

33 Blagdon, *Modern Discoveries*, Vol. 1, pp. xlii–xliii.

34 Blagdon, *Modern Discoveries*, Vol. 1, pp. xliv–xlv.

35 Pinkerton, *Modern Geography*. Pinkerton also later compiled *A General Collection of the best and most interesting voyages and travels*.

36 *The Imperial Review; or, London and Dublin Literary Journal* **2** (1804), p. 326.

37 Goldsmith [Sir Richard Phillips], *General View*, Vol. 1, pp. xiii, xiv, xviii respectively.
38 Goldsmith, *General View*, Vol. 1, pp. 43–4.
39 Goldsmith, *General View*, Vol. 1, p. 46.
40 Shaffer, *'Kubla Khan' and the Fall of Jerusalem*, chapter 3.
41 Preston, 'Biblical Criticism, Literature, and the Eighteenth-century Reader', in Rivers, ed., *Books and their Readers*, p. 98.
42 Preston, 'Biblical Criticism', p. 113; Harris, 'Allegory to analogy'.
43 Biographer's comment in Otter, *Remains*, Vol. 1, pp. 446–7; Clarke to Otter, in *ibid.*, Vol. 2, pp. 106–8.
44 Clarke's religious orientation is quite difficult to assess, although his involvement with the Unitarian community warrants further examination. Suffice it to say here that his Jesus College tutor was William Frend, who was expelled from the University in 1793 for his Unitarian beliefs. Clarke remained close with Frend. Also, John Tweddell, coeval traveller with Clarke and an acquaintance, who died while in Athens in 1799, was a defender of the French Revolution and tightly bound with the circle at Jesus College connecting Frend, Samuel Coleridge and Robert Tyrwhit, also a friend of Clarke's. For religion at Cambridge, see John Gascoigne, *Cambridge*, Young, *Religion and Enlightenment*.
45 Fowler, *The Eastern Mirror*, pp. iv–v. The references include Harmer's 4-volume *Observations on Various passages of Scripture* and Calmet's 2 volume *Dictionary of the Bible*.
46 Fowler, *The Eastern Mirror*, p. 38.
47 H.D. Whittington to Lord Aberdeen, 27 January 1805, quoted in Liscombe, *William Wilkins*, p. 44 for first quotation; Otter, *Remains*, Vol. 2, pp. 209–210 for latter quotation.
48 Venn, *Biographical History of Gonville and Caius College*, Vol. 2, pp. 114, 134.
49 As with H.D. Whittington; see Otter, *Remains*, Vol. 2, pp. 198–200.
50 Material of this sort is found in Cambridge University Library, University Archives, UP.1; individual expenses found in CUL UA CUP 20/1. See Leedham-Green, 'University Press Records'.
51 Clarke to Otter, in Otter, *Remains*, Vol. 2, p. 234. See also, Gascoigne, *Cambridge*, p. 295; Porter, *The Making of Geology*, p. 144; Becher, 'Voluntary Science in Nineteenth Century Cambridge'.
52 An account of the cork model found in Clarke to Miss Newling, 2 January 1814, CUL MSS ADD 7082, f. 75; Hamilton's alleged comment in Otter, *Remains*, Vol. 1, p. 126.
53 Quoted in *Cambridge University Calendar* (Cambridge, 1809).
54 Wright, *Alma Mater*, Vol. 2, pp. 30–1.
55 [Gooch], *Facetiae Cantabrigienses*, pp. 150–2.
56 A more detailed account of the politics of Clarke's professorship is offered in Dolan, *Governing Matters*, chapter 2.
57 Monk and Blomfield, eds, *Museum Criticum; or, Cambridge Classical Researches*, 7 nos (Cambridge, 1814–26), No. 1, p. v and No. 6, p. 128, respectively.
58 Watson, *Richard Porson*, p. 367.
59 *Edinburgh Review* **15** (1810), 453–8, p. 453.

60 Clarke, *Travels*, Vol. 4 (first edition), p. xii.
61 Clarke, *Travels*, Vol. 4 (first edition), p. xii.
62 By 1800, 190 editions of Pliny's work had been published: see Gudger, 'Pliny's Historia naturalis'.
63 Clarke, *Travels*, Vol. 2 (first edition), p. 336.
64 Clarke, *Travels*, Vol. 2 (first edition), p. 384.
65 Warburton's annotated syllabus, pp. 31, 501.
66 Clarke, *Travels*, Vol. 2 (first edition), p. 384.
67 Notes describing what he collected for his lectures or for donation are scattered throughout his *Travels*, including those above at: Vol. III, pp. 13, 25, 197, 200; Vol. VI, pp. 129, 213, 219; Vol. IX, pp. 149, 319, 419, 463; Vol. X, pp. 4, 15 (all references to fourth edition).
68 Walker, *An Essay on the Education of the People*, p. 5.
69 On excessive concerns for proper preparation, see the 500-page list of questions travellers should seek answers to in Berchtold, *Patriotic Travellers*, Vol. 1.
70 Kelsall, *Phantasm*, p. 14.
71 Kelsall, *Phantasm*, pp. 18–24.
72 Kelsall, *Phantasm*, pp. 38, 45.
73 Kelsall, *Phantasm*, pp. 52–3.
74 Humboldt, 'Intellectual Institutions in Berlin'.
75 Humboldt, 'Intellectual Institutions in Berlin', pp. 242, 246, 245 respectively.
76 See Shaffer, 'Romantic philosophy', in Cunningham and Jardine, eds, *Romanticism and the Sciences*; Ziolkowski, *German Romanticism*, pp. 286–94.
77 Clarke, 'Preface' to his *Travels*, Vol. 1 (second edition), on Kelsall's assistance. For Kelsall's travels, see Kelsall, *A Letter from Athens*; idem, *Classical Excursion*. While not explicit here, Humboldt's ideas about the 'universe of knowledge' and disciplinary studies did not neglect the benefits of scientific travel, which were, of course, clearly demonstrated through the extensive travels of his brother, Alexander von Humboldt.
78 Lady Spencer to Lady Bessborough, 22 July 1804, in Bessborough, ed., *Lady Bessborough*, pp. 126–7.
79 Macvey Napier[?], reviewing J. Griffiths, *Travels*, in *Edinburgh Review* 8 (1806), 35–51, p. 35.
80 Ibid., p. 36.
81 Woolf, 'The Construction of a European World-View,' p. 74.
82 Woolf, 'The Construction of a European World-View,' pp. 77–8; emphasis added.
83 Marshall and Williams, *Great Map of Mankind*, pp. 92–3.
84 Kiernan, *The Lords of Human Kind*.
85 See Gascoigne, *Joseph Banks*, chapter 4, 'From Antiquarian to Anthropologist' for a good overview of emerging theories in eighteenth-century anthropology.
86 Quoted in Marshall and Williams, *Great Map of Mankind*, p. 92.
87 Quoted in Batten, *Pleasurable Instructions*, p. 7.
88 Home (Lord Kames), *Sketches on the History of Man*, Vol. 1, p. 70.
89 Marsden, *History of Sumatra*, pp. 169–70.
90 Smith, *Causes of the Variety of Complexion*, pp. 62–3.

91 Smith, *Causes of Variety of Complexion*, pp. 68–70.
92 Home (Lord Kames), *Sketches on the History of Man*, Vol. 1, p. 25.
93 Marshall and Williams, *Great Map of Mankind*, p. 88.
94 Quoted in Febvre, 'Civilisation', in Burke, ed., *A New Kind of History*, p. 222.
95 The counter-position – the assertion of a 'French civilisation' – is clearly demonstrated, for example, in two courses taught by François Guizot at the Sorbonne in 1828 and 1829: 'La Civilisation en Europe' and 'La Civilisation en France', respectively. In the former, Guizot sketched the condition of civilisations in Europe. Febvre summarised the lecture: 'So a rapid review of all the various European civilisations was sufficient to show him in England a civilisation almost exclusively orientated towards social perfection but whose representatives proved to be lacking in the talent required "to light those great intellectual torches which illuminate whole eras." Conversely, German civilisation was powerful in its spirit but feeble in its organisation and in its attainment of social perfection. . . . On the other hand, there was a country, the only one, able to pursue the harmonious development of ideas and facts, of the intellectual and material order – that country was of course France . . .' Febvre, in Burke, ed., *A New Kind of History*, p. 242.
96 Coxe, *Travels*, quoted in Cross, *By the Banks of the Neva*, p. 352.
97 For the ways in which frontier lands and cartographical boundaries also helped shape the 'patriotic sentiment of the Enlightenment', see Evans, 'Frontiers and National Identities'.
98 Richardson, *A Dissertation*, p. 125.

Bibliography

Adams, Percy. *Travellers and Travel Liars, 1660–1800* (London: Constable, 1980).

Alexander, John T. *Emperor of the Cossacks: Pugachev and the Frontier Jacquerie of 1773–1775* (Lawrence, Kansas: Coronado Press, 1973).

Anderson, M.S. *Britain's Discovery of Russia, 1553–1815* (London: Macmillan, 1958).

Anderson, M.S. *Peter the Great* (2nd edn, London and New York: Longman, 1995).

Andersson, Björn. 'Runes, Magic, and Ideology: Defining a Discipline', in Brian Dolan, ed., *Science Unbound: Geography, Space & Discipline* (Umeå: Umeå University Press, 1998).

Angelomatis-Tsougarakis, Helen. *The Eve of the Greek Revival: British Travellers 'Perceptions of Early Nineteenth-Century Greece* (London and New York: Routledge, 1990).

Anon. 'On the British Museum, and on Collectors', *The Quarterly Journal of Science and the Arts* 7 (1819), 259–66.

Anon. *Description of the House and Museum on the north side of Lincoln's Inn Fields, the residence of John Soane* (London, 1830).

Anon. *The Danger of the Political Balance of Europe, translated from the French of the King of Sweden, by the Rt. Hon. Lord Mountmorres* (frequently attributed to Gustav III of Sweden) (London, 1790).

Appleby, John. 'British Doctors in Russia, 1657–1807: Their Contribution to Anglo-Russian Medical and Natural History', PhD, University of East Anglia, 1979.

Astley, Thomas. *A New General Collection of Voyages and Travels*, 5 vols (London, 1745–7).

Augustinos, Olga. *French Odysseys: Greece in French Travel Literature from the Renaissance to the Romantic Era* (Baltimore and London: Johns Hopkins University Press, 1994).

Avery, John. *Progress, Poverty, and Population: Re-reading Condorcet, Godwin, and Malthus* (Essex: Frank Cass, 1997).

Baker, Keith Michael. *Condorcet: From Natural Philosophy to Social Mathematics* (Chicago: University of Chicago Press, 1975, 2nd edn, 1982).

Banks, R.E.R. et al., eds. *Sir Joseph Banks: A Global Perspective* (Kew: Royal Botanic Gardens, 1994).

Bartlett, Roger and Janet M. Hartley, eds. *Russia in the Age of Enlightenment: Essays for Isabel de Madariaga* (London: Macmillan, 1990).

Barton, H.A. 'Late Gustavian Autocracy in Sweden', *Scandinavian Studies* **46** (1974), 280–95.

Bassin, Mark. 'Russia between Europe and Asia: The Ideological Construction of Geographical Space', *Slavic Review* **50** (1991), 1–16.

Batten, Charles. 'Literary Responses to the Eighteenth-Century Voyages', in D. Howse, ed., *Background to Discovery: Pacific Exploration from Dampier to Cook* (Berkeley: University of California Press, 1990), 128–59.

Batten, Charles. *Pleasurable Instruction: Form and Convention in Eighteenth-century Travel Literature* (Berkeley: University of California Press, 1978).

Bayly, C.A. *Imperial Meridian: The British Empire and the World, 1780–1830* (London and New York: Longman, 1989).

Beaglehole, J.C. 'Eighteenth-Century Science and the Voyages of Discovery', *New Zealand Journal of History* **3** (1969), 107–23.

Beaucour, F., Y. Laissus and C. Orgogozo, *The Discovery of Egypt: Artists, Travellers, and Scientists* (Paris: Flammarion, 1990).

Becher, H. 'Voluntary Science in Nineteenth Century Cambridge University to the 1850s', *British Journal for the History of Science* **19** (1986), 57–87.

Beer, Gillian. *The Romance* (London: Methuen, 1970).

Bell, John. *Travels from St Petersburg in Russia, to Diverse Parts of Asia*, 2 vols (Glasgow, 1763).

Berchtold, Leopold. *An Essay to direct and extend the Inquiries of Patriotic Travellers*, 2 vols (London, 1789).

Berman, Morris. *Social Change and Scientific Organization: The Royal Institution, 1799–1844* (London: Heinemann, 1978).

Bernal, Martin. *'Black Athena: The Afro-Asiatic Roots of Classical Civilisation: Volume I: The Fabrication of Ancient Greece 1785–1985* (London: Vintage, 1991).

Bessborough, Earl of, ed. *Lady Bessborough and her Family Circle* (London: John Murray, 1940).

Besterman, T. *The Publishing Firm of Cadell and Davies: Select Correspondence and Accounts, 1793–1836* (Oxford: Oxford University Press, 1938).

Black, Jeremy. *Maps and History: Constructing Images of the Past* (New Haven and London: Yale University Press, 1997).

Black, Jeremy. *The British Abroad: The Grand Tour in the Eighteenth Century* (New York: St. Martin's Press, 1992).

Blagdon, Francis. *Modern Discoveries; or, a Collection of Facts and Observations, principally relative to the various Branches of Natural History, resulting from the Geological, Topographical, Botanical, Physiological, Mineralogical, and Philosophical Researches of Celebrated Modern Travellers in every quarter of the Globe*, 8 vols (London, 1802–3).

Bonar, James. *Malthus and his Work* (London: Frank Cass, 1885; reprinted 1966).

Boswell, James. *The Life of Samuel Johnson*, 2 vols (London 1791).

Brändström, Anders and Lars-Göran Tedebrand, eds, *Society, Health and Population during the Demographic Transition* (Stockholm: Almqvist and Wiksell, 1988).

Brewer, John. *Pleasures of the Imagination: English Culture in the Eighteenth Century* (London: HarperCollins, 1997).

Briggs, Asa. *The Age of Improvement, 1783–1867* (London and New York: Longman, 1959, reprint 1993).

Broberg, Gunnar. 'Homo sapiens: Linnaeus's Classification of Man', in Tore Frängsmyr, ed., *Linnaeus: The Man and His Work* (Canton, MA: Science History Publications, 1994), 156–94.

Brock, M.G. and M.C. Curthoys, eds. *The History of the University of Oxford: Volume VI: Nineteenth-century Oxford, Part I* (Oxford: Clarendon Press, 1997).

Bullock, William. *A Companion to the Liverpool Museum, Containing a brief*

Description of upwards of Four Thousand of its natural & Foreign Curiosities, Antiquities, & Productions of the Fine Arts (6th edn, Hull, 1808).

Burke, Peter. *The Renaissance Sense of the Past* (New York: St. Martin's Press, 1970).

Burrow, J.W. *A Liberal Descent: Victorian Historians and the English Past* (Cambridge: Cambridge University Press, 1981).

Butler, Marilyn. 'Romanticism in England', in R. Porter and M. Teich, eds, *Romanticism in National Context* (Cambridge: Cambridge University Press, 1988), 37–67.

Buzard, James. *The Beaten Track: European Tourism, Literature, and the Ways to 'Culture', 1800–1918* (Oxford: Clarendon Press, 1993).

Bynum, W.F. and R. Porter, eds. *Companion Encyclopaedia of the History of Medicine* (London: Routledge, 1993).

Calmet, Augustin. *Dictionary of the Bible*, 2 vols (Augustae Vindelicorum, 1776).

[Campbell, John] *The Polite Correspondence: or, Rational Amusement* (London, 1740).

Campbell, John, ed. *Navigantium atque Itinerantium Bibliotheca; or, A Compleate Collection of Voyages and Travels* (London, 1744).

Campbell, Mary. *The Witness and the Other World: Exotic European Travel Writing, 400–1600* (Ithaca, NY: Cornell University Press, 1988).

Cantarel-Besson, Yveline. *La naissance du Musée du Louvre, la politique muséologique sous la Révolution d àpres, les archives des musées nationaux* (Paris: Editions de la Réunion des musées nationaux, 1981).

Carr, John. *A Northern Summer: or, Travels round the Baltic, through Denmark, Sweden, Russia, Prussia, and Part of Germany, in the year 1804* (London: for Richard Phillips, 1805).

Carr, John. *A Northern Summer: or, Travels round the Baltic, through Denmark, Sweden, Russia, and Part of Germany, in the year 1804* (London, 1805).

Caygill, M. *The Story of the British Museum* (London: British Museum Press, 1981).

Chandler, Richard. *Travels in Asia Minor* (Oxford, 1775; second edition, London, 1776).

Chandler, Richard. *Travels in Asia Minor, and Greece; or, an Account of a Tour made at the Expense of the Society of Dilettanti*, 2 vols (in one) (London, 1817, reprint 1825).

Chandler, Richard. *Travels in Greece* (Oxford, 1776).

Christie, Ian R. *The Benthams in Russia: 1780–1791* (Oxford: Berg, 1993).

Clarke, Edward Daniel. *A Tour through the South of England, Wales, and Part of Ireland, Made During the Summer of 1791* (London, 1793).

Clarke, Edward Daniel. *Testimonies of different authors respecting the Colossal Statue of Ceres* (Cambridge, 1802).

Clarke, Edward Daniel. *The Tomb of Alexander, a dissertation on the Sarcophagus brought from Alexandria, and now in the British Museum* (Cambridge, 1805).

Clarke, Edward Daniel. *Greek Marbles brought from the shores of the Euxine, Archipelago, and Mediterranean, and deposited in the Vestibule of the University of Cambridge* (Cambridge, 1809).

Clarke, Edward Daniel. *Travels in Various Countries of Europe, Asia, and Africa*, 1st edn, 6 vols (London, 1810–23).

Clarke, Edward Daniel. *Travels in Various Countries of Europe, Asia, and Africa*, 3rd edn, 11 vols (London, 1816–24).

Clarke, Edward Daniel. *Critique on the Character and Writings of Sir George Wheler, as a traveller* (York, 1820).

Cohen, Michèle. 'The Grand Tour: Constructing the English Gentleman in Eighteenth-century France', *History of Education* **21** (1992), 241–57.

Colley, Linda. *Britons: Forging the Nation, 1707–1837* (New Haven and London: Yale University Press, 1992).

Collier, William. *A Tour of Russia, Siberia, and the Crimea 1792–1794* (London: Frank Cass, 1971).

Collinder, Björn. *The Lapps* (Princeton, NJ: Princeton University Press, 1949).

Collins, A.S. *The Profession of Letters: a Study of the Relation of Author to Patron, Publisher, and Public, 1780–1832* (London: George Routledge & Sons, 1928).

Connelly, O. *The French Revolution and Napoleonic Era*, 2nd edn (Florida: Holt, Rinehart, and Winston, 1991).

Consett, Matthew. *A Tour through Sweden, Swedish-Lapland, Finland and Denmark* (London, 1789).

Constantine, David. *Early Greek Travellers and the Hellenic Ideal* (Cambridge: Cambridge University Press, 1984).

Cook, B.F. *The Elgin Marbles* (London: British Museum Publications, 1993).

Couloubaritsis, L., et al. *The Origins of European Identity* (Brussels: European Interuniversity Press, 1993).

Coxe, William. *Travels into Poland, Russia, Sweden, and Denmark, interspersed with historical relations and political inquiries*, 3 vols (London: T. Cadell, 1784).

Cronin, Vincent. *Catherine: Empress of all the Russias* (London: Collins, 1978).

Crook, J.M. *The Greek Revival: Neo-classical Attitudes in British Architecture 1760–1870* (London: John Murray, 1972).

Cross, Anthony. 'The Reverend William Tooke's Contributions to English Knowledge of Russia at the End of the Eighteenth Century', *Canadian Slavic Studies* **3** (1969), 106–15.

Cross, Anthony. 'Chaplains to the British Factory in St Petersburg, 1723–1813', *European Studies Review* **2** (1972), 125–42.

Cross, Anthony, ed. *Great Britain and Russia in the Eighteenth Century: Contacts and Comparisons* (Newtonville, MA: Oriental Research Partners, 1979).

Cross, Anthony. 'British Knowledge of Russian Culture (1698–1801)', *Canadian-American Slavic Studies* **13** (1979), 412–35.

Cross, Anthony. 'The British in Catherine's Russia: A Preliminary Survey', in Garrard, ed., *Eighteenth Century in Russia*, 233–63.

Cross, Anthony. *Anglo-Russica: Aspects of Cultural Relations between Great Britain and Russia in the Eighteenth and Early Nineteenth Centuries* (Oxford: Berg, 1993).

Cross, Anthony. 'Sir John Sinclair's *General Observations* on Russia', in Cross, *Anglo-Russica*, 51–61.

Cross, Anthony. *By the Banks of the Neva: Chapters from the Lives and Careers of the British in Eighteenth-century Russia* (Cambridge: Cambridge University Press, 1997).

Cunningham, Andrew and Nicholas Jardine, eds. *Romanticism and the Sciences* (Cambridge: Cambridge University Press, 1990).

Cunningham, Andrew and Perry Williams, 'De-centring the "big picture":
The origins of modern science and the modern origins of science', *British
Journal for the History of Science* **26** (1993), 407–32.

Cunningham, J.W. *Cautions to Continental Travellers* (Cambridge, 1818).

Cust, Lionel and Sidney Coluin. *History of the Society of Dilettanti* (London,
1898).

Dallaway, James. *Anecdotes of the Arts in England: or, Comparative remarks on
architecture, sculpture, and painting chiefly illustrated by specimens at Oxford*
(London, 1800).

Dallaway, James. *Constantinople Ancient and Modern with Excursions to the
Archipelago and to the Troad* (London, 1797).

Darnton, Robert. *The Business of Enlightenment: a Publishing History of the
'Encyclopédie,' 1775–1800* (London: Belknap Press, 1979).

De Beer, G.R. *Sir Hans Sloane and the British Museum* (London, 1953).

De Bruin, Cornelis. *Voyage to the Levant and Travels into Muscovey*, 3 vols
(London, 1720, with later editions in 1737 and 1759).

De Madariaga, Isabel. *Britain, Russia, and the Armed Neutrality of 1780: Sir
James Harris's Mission to St Petersburg during the American Revolution* (New
Haven, CT: Yale University Press, 1962).

De Madariaga, Isabel. *Russia in the Age of Catherine the Great* (London:
Weidenfeld and Nicolson, 1981).

De Madariaga, Isabel. 'The Russian Nobility in the Seventeenth and Eighteenth Centuries', in H.M. Scott, ed., *The European Nobilities in the Seventeenth
and Eighteenth Centuries* (London and New York: Longman, 1995), Vol. 2,
chapter 8.

Delanty, Gerard. *Inventing Europe: Idea, Identity, Reality* (London: Macmillan,
1995).

Dening, Greg. *The Death of William Gooch: A History's Anthropology* (Hono-
lulu: University of Hawaii Press, 1995).

Desmond, Adrian and James Moore, *Darwin* (London: Penguin, 1992).

Dick, P. *A Descriptive Catalogue of a Museum of Antiquities and Foreign Curiosities, Natural and Artificial; including Models illustrative of Military and
Naval Affairs, Armour and Weapons, Instruments of Torture, Polytheism, Sep-
ulchres, with the Manner of Depositing the Dead; the costume of different
nations, manuscripts, natural history, including anatomy &c* (London, 1815).

Digby, Anne. 'Malthus and the Reform of the English Poor Law', in
Michael Turner, ed., *Malthus and his Time* (Basingstoke: Macmillan, 1986),
157–69.

Dmytryshyn, Basil, ed. *The Modernisation of Russia under Peter I and Catherine
II* (New York, 1974).

Dodwell, Edward. *Classical and Topographical Tour through Greece, during the
years 1801, 1805, and 1806*, 2 vols (London, 1819).

Dolan, Brian. *Governing Matters: The Values of English Education in the Earth
Sciences, 1790–1830* (PhD dissertation, Cambridge University, 1995).

Dolan, Brian, ed. *Science Unbound: Geography, Space & Discipline* (Umeå: Umeå
University Press, 1998).

Dolan, Brian. 'Transferring Skill: Blowpipe Analysis in Sweden in Britain,
1750–1850', in idem., ed., *Science Unbound: Geography, Space & Discipline*
(Umeå: Umeå University Press, 1998).

Dolan, Brian, ed. *Malthus, Medicine and Morality: 'Malthusianism' after 1798* (Amsterdam and Atlanta, GA: Rodopi, 1999).

Douthwaite, J. 'Rewriting the Savage: The Extraordinary Fictions of the "Wild Girl of Champagne" ', *Eighteenth-century Studies* **28** (1994–5), 163–92.

Drake, Michael. 'Malthus on Norway', *Population Studies: A Journal of Demography* **20** (1966), 175–96.

Dyer, G. *The Privileges of the University of Cambridge*, 2 vols (London, 1824).

Elias, Norbert. *The Civilizing Process: (I) The History of Manners, (II) State Formation and Civilization* (London: Basil Blackwell, 1978).

Ella, Anthony. *Visits to the Leverian Museum; containing an account of several of its principal curiosities, &c* (London, 1805).

Emerson, Ralph Waldo. *Conduct of Life* (London, 1860).

Emsley, C. *The Longman Companion to Napoleonic Europe* (New York and London: Longman, 1993).

Engeström, Gustav von. *Guide du Voyages* (Stockholm, 1797).

Evans, R.J.W. 'Frontiers and National Identities in Central Europe', *International History Review* **14** (1992), 480–502.

Febvre, Lucien. Civilisation: Evolution of a Word and Group of Ideas', in Peter Burke, ed., *A New Kind of History: From the Writings of Febvre*, trans. K. Folca (London: Routledge and Kegan Paul, 1973), 219–57.

Fedorov, A.S. 'Russia and Britain in the Eighteenth Century: A Survey of Economic and Scientific Links', in A. Cross, ed., *Great Britain and Russia in the Eighteenth Century: Contacts and Comparisons* (Newtonville, MA: Oriental Research Partners, 1979), 137–44.

Ferguson, Adam. *An Essay on the History of Civil Society* (London, 1767; London: Transaction Books, 1980).

Ferguson, Adam. *Principles of Moral and Political Science* (London, 1792).

Firestone, Clark B. *The Coasts of Illusion: A Study of Travel Tales* (New York and London: Harper & Brothers, 1924).

Fisher, A.W. *Crimean Tatars* (Stanford: University of California Press, 1978).

Fisher, R. and H. Johnston, eds. *Captain James Cook and His Times* (London: Croom Helm, 1979).

Fowler, W. *The Eastern Mirror; An Illustration of the Sacred Scriptures; in which the Customs of the Oriental Nations are clearly Developed by the Writings of the Most Celebrated Travellers* (Exeter, 1814).

Fox, Christopher, Roy Porter and Robert Wokler, eds, *Inventing Human Science: Eighteenth-Century Domains* (Berkeley and London: University of California Press, 1995).

Frängsmyr, Tore, ed. *Linnaeus: The Man and His Work* (Canton, MA: Science History Publications, 1994).

Frängsmyr, Tore, ed. *Science in Sweden: The Royal Swedish Academy of Sciences 1739–1989* (Canton, MA: Science History Publications, 1989).

Frantz, R.W. *The English Traveller and the Movement of Ideas 1660–1732* (New York, 1968).

Fraser, N. *Continental Drifts: Travels in the New Europe* (London: Secker and Warburg, 1997).

Friis, H.R., ed. *The Pacific Basin: A History of its Geographical Exploration* (New York: American Geographical Society, 1967).

Gaisford, D. *Catalogus, sive Notitia Manuscriptorum quae a cel. E.D.C. comparata in Bibliotecheca Bodleiana adservantur, &c.* (Oxford, 1812).

Garrard, J.G., ed. *The Eighteenth Century in Russia* (Oxford: Clarendon Press, 1973).

Gascoigne, John. *Cambridge in the Age of the Enlightenment: Science, Religion, and Politics from the Restoration to the French Revolution* (Cambridge: Cambridge University Press, 1989).

Gascoigne, John. *Joseph Banks and the English Enlightenment: Useful Knowledge and Polite Culture* (Cambridge: Cambridge University Press, 1994).

Gell, William. *Topography of Troy* (London, 1804).

Gell, William. *Geography and Antiquities of Ithaca* (London, 1807).

Gell, William. *Itinerary of Greece* (London, 1810).

Gell, William. *Itinerary of Morea* (London, 1817).

Gell, William. *Pompeiana*, 2 vols (London, 1817–19).

Gell, William. *Journey in the Morea* (London, 1823).

Gell, William. *Topography of Rome* (London, 1834).

Gibbon, Edward. *The Decline and Fall of the Roman Empire*, 7 vols (London, 1896–1900).

Gigerenzer, Gerd, Zeno Swijtink, Theodore Porter, Lorraine Daston, John Beatty and Lorenz Krüger. *The Empire of Chance: How Probability Changed Science and Everyday Life* (Cambridge: Cambridge University Press, 1989).

Gmelin, Johann Georg. *Flora Sibirica, sive historia plantarum Sibiriae*, 4 vols (Petropoli, 1747–69).

Goldsmith, J. [Sir Richard Phillips], *A General View of the Manners, Customs, and Curiosities of Different Nations; including a Geographical Description of the Earth*, 2 vols (Philadelphia, 1818).

Goldsmith, Oliver. *The Citizen of the World* (London, 1762; London: Dent, 1934).

[Gooch, R.] *Facetiae Cantabrigienses: Consisting of Anecdotes, Smart Sayings, Satirics, Retorts, &c* (London: Charles Mason, 1836).

Graham, P., ed. *Byron's Bulldog: The Letters of John Cam Hobhouse to Lord Byron* (Columbus, OH: Ohio State University Press, 1984).

Greene, John. *The Death of Adam: Evolution and its Impact on Western Thought* (Ames, Iowa: Iowa State University Press, 1959).

Greengrass, Mark, ed. *Conquest and Coalescence: the Shaping of the State in Early Modern Europe* (London: Edward Arnold, 1991).

Grey, Ian. *Catherine the Great: Autocrat and Empress of all Russia* (London: Hodder & Stoughton, 1961).

Griffiths, J. *Travels in Europe, Asia Minor, and Arabia* (London, 1805).

Gudger, E.W. 'Pliny's Historia naturalis: The Most Popular Natural History Ever Published', *Isis* **6** (1924), 269–81.

Guthrie, Maria. *A Tour performed in the years 1795–1796, through the Taurida, or Crimea* (London: Cadell and Davies, 1802).

Habermas, J. *The Structural Transformation of the Public Sphere*, trans. Thomas Burger (Cambridge, MA: MIT Press, 1989).

Hacking, Ian. *The Taming of Chance* (Cambridge: Cambridge University Press, 1990).

Hallendorff, Carl and Adolf Schück, *History of Sweden* (London: Cassell and Company, Ltd., 1929).

Hamilton, Horace Ernst. *Travel and Science in Thomson's 'Seasons* (PhD dissertation, Yale University, 1941).

Hamilton, William R. *Memorandum on the Subject of the Earl of Elgin's Pursuits in Greece* (Edinburgh, 1810; new edn London, 1811; 2nd edn London, 1815).

Hans, N. *New Trends in Education in the Eighteenth Century*, 2nd edn (London: Routledge & Kegan Paul, 1966).

Hanway, Jonas. *An Historical Account of British Trade over the Caspian: with the Author's Journal of Travels from England through Russia into Persia* (London, 1753).

Hargreaves, A.G. 'European Identity and the Colonial Frontier', *Journal of European Studies* **12** (1982), 66–79.

Harmer, Thomas. *Observations on Various passages of Scripture, placing them in a new light . . . from relations in books of voyages and travels in the East*, 4 vols (London, 1776–87).

Harraden, R. *Cantabrigia Depicta: A Series of Engravings representing the Most Picturesque and Interesting Edifices in the University of Cambridge, with an Historical and Descriptive Account of Each* (Cambridge, 1809).

Harris, V. 'Allegory to Analogy in the Interpretation of Scriptures', *Philological Quarterly* **45** (1966), 1–23.

Haskell, F. and N. Penny. *Taste and the Antique: The Lure of Classical Sculpture 1500–1900* (New Haven and London: Yale University Press, 1981).

Hatfield, H.C. *Winckelmann and his German Critics* (New York, 1943).

Haynes, D.E.L. *The Arundel Marbles* (Oxford: Ashmolean Museum, 1975).

Herder, J.G. von. *Outlines of a Philosophy of the History of Man*, trans. T. Churchill (London, 1800).

Heriot, George. *Travels through the Canadas* (London, 1807).

Hibbert, Christopher. *The Grand Tour* (London: Methuen, 1987).

Hilles, F.W. and H. Bloom, eds. *From Sensibility to Romanticism* (New York: Oxford University Press, 1965).

Hilton, Boyd. *The Age of Atonement: The Influence of Evangelicalism on Social and Economic Thought* (Oxford: Clarendon Press, 1988).

Hitchens, Christopher. *The Elgin Marbles: Should they be Returned to Greece?* (London: Chatto & Windus, 1987; Verso, 1997).

Hobhouse, John Cam. *A Journey through Albania and Other Provinces of Turkey in Europe and Asia, to Constantinople During the years 1809 and 1810*, 2 vols (London: James Cawthorn, 1813).

Holland, Henry. *Travels in the Ionian Isles . . during the years 1812 and 1813* (London, 1815; 2nd edn, 2 vols, 1819).

Home, Henry (Lord Kames). *Sketches on the History of Man*, 4 vols (1778, facsimile reprint London: Routledge, 1993).

Hont, Istvan and Michael Ignatieff, eds. *Wealth and Virtue: The Shaping of Political Economy in the Scottish Enlightenment* (Cambridge: Cambridge University Press, 1983).

Hosking, Geoffrey. *Russia: People and Empire, 1552–1917* (London: HarperCollins, 1997).

Hudson, Nicholas. *Writing and European Thought 1600–1830* (Cambridge: Cambridge University Press, 1994).

Hughes, Lindsay. *Russia under Peter the Great* (New Haven, CT: Yale University Press, 1998).

Hughes, Thomas Smart. *Travels in Greece and Albania*, 2 vols (London, 1820).

Humboldt, W. von. 'On the Spirit and Organisational Framework of Intellectual Institutions in Berlin', translated and abridged in 'University Reform in Germany', *Minerva* **8** (1970), 242–50.

Hume, David. *Essays: Moral, Political, and Literary*, Eugene F. Miller, ed. (Indianapolis: Liberty Classics, 1987).

Hunt, Lynn. *Politics, Culture, and Class in the French Revolution* (London: Methuen, 1986).

Iliffe, Rob. ' "Aplatisseur du Monde et de Cassini": Maupertuis, Precision Measurement, and the Shape of the Earth in the 1730s', *History of Science* **31** (1993), 335–75.

Jacyna, L.S. *Philosophic Whigs: Medicine, Science and Citizenship in Edinburgh, 1789–1848* (London and New York: Routledge, 1994).

James, Patricia. *The Travel Diaries of Thomas Robert Malthus* (Cambridge: Cambridge University Press, 1966).

Jenkins, Ian. *Archaeologists and Aesthetes in the Sculpture Galleries of the British Museum, 1800–1939* (London: British Museum, 1992).

Johannisson, Karin. 'Naturvetenskap på reträtt: En diskussion om naturvetenskapens status under svenskt 1700-tal,' *Lychnos* (1979–80), 109–53.

Johannisson, Karin. 'Why Cure the Sick? Population Policy and Health Programs within 18th-Century Swedish Mercantilism', in Anders Brändström and Lars-Göran Tedebrand, eds, *Society, Health and Population during the Demographic Transition* (Stockholm: Almqvist and Wiksell, 1988), 323–30.

Johannisson, Karin. 'The People's Health: Public Health Policies in Sweden', in Dorothy Porter, ed., *The History of Public Health and the Modern State* (Amsterdam: Rodopi, 1994), 165–82.

Jones, R.F. *Ancients and Moderns: A Study of the Background of the Battle of the Books* (St. Louis: Washington University Press, 1936).

Jones, Robert E. 'Urban Planning and the Development of Provincial Towns in Russia, 1762–1796', in J.G. Garrard, ed., *The Eighteenth Century in Russia* (Oxford: Clarendon Press, 1973), 321–44.

Kelsall, Charles. *A Letter from Athens to a Friend in England* (London, 1813).

Kelsall, Charles. *Phantasm of an University: With Prolegomena* (London, 1814).

Kelsall, Charles. *Classical Excursion from Rome to Arpino, Geneva* (London, 1820).

Kennedy, Emmet. *A Cultural History of the French Revolution* (New Haven and London: Yale University Press, 1989).

Kennedy, James J., Jr. 'The Politics of Assassination', in Hugh Ragsdale, ed., *Paul I: A Reassessment of His Life and Reign* (Pittsburgh, PA: UCIS Series in Russian and East European Studies, 1979), 125–45.

Kent, H.S. *War and Trade in Northern Seas: Anglo-Scandinavian Economic Relations in the Mid-Eighteenth Century* (Cambridge: Cambridge University Press, 1973).

Khodarkovsky, Michael. 'From Frontier to Empire: The Concept of the Frontier in Russia, Sixteenth–Eighteenth Centuries', *Russian History* **19** (1992), 115–28.

Kiernan, V.G. *The Lords of Human Kind: European attitudes towards the outside world in the Imperial Age* (London: Weidenfeld and Nicolson, 1969).

Kingston-Mann, Esther. *In Search of the True West: Culture, Economics, and Problems of Russian Development* (Princeton, NJ: Princeton University Press, 1998).

Kirby, D. *Northern Europe in the Early Modern Period: The Baltic World, 1492–1792* (London: Longman, 1990).

Koerner, Lisbet. 'Purposes of Linnaean Travel: A Preliminary Research Report', in D. Miller and P. Reill, eds, *Visions of Empire: Voyages, Botany, and Representations of Nature* (Cambridge: Cambridge University Press, 1996), 117–52.

Kohut, Zenon. 'Ukraine: From Autonomy to Integration (1654–1830s)', in Mark Greengrass, ed., *Conquest and Coalescence: the Shaping of the State in Early Modern Europe* (London: Edward Arnold, 1991).

Kors, A.C. and Paul Korshin, eds. *Anticipation of the Enlightenment in England, France, and Germany* (Philadelphia: University of Pennsylvania Press, 1987).

Kriegel, A.D. 'Liberty and Whiggery in Early Nineteenth-century England', *Journal of Modern History* **52** (1980), 253–78.

Labbe, Jacqueline. 'A Family Romance: Mary Wollstonecraft, Mary Godwin, and Travel', *Genre* **25** (1992), 211–28.

Lawson, Philip. *The East India Company: A History* (London and New York: Longman, 1993).

Leask, Nigel. *British Romantic Writers and the East: Anxieties of Empire* (Cambridge: Cambridge University Press, 1992).

Leedham-Green, E. 'University Press Records in the University Archives: An Account and a Checklist', *Transactions of the Cambridge Bibliographical Society* **8** (1984), 398–418.

Lefebvre, G. *The French Revolution from 1739 to 1799*, trans. J.H. Stewart and J. Friguglietti (London: Routledge and Kegan Paul, 1964).

Leith, James. *The Idea of Art as Propaganda in France, 1750–1799* (Toronto, 1965).

Leoussi, Athena. *Nationalism and Classicism: The Classical Body as National Symbol in Nineteenth-century England and France* (London: Macmillan, 1998).

Leppmann, W. *Winckelmann* (London, 1971).

Levine, Joseph. *The Battle of the Books: History and Literature in the Augustan Age* (Ithaca, NY: Cornell University Press, 1991).

Lindqvist, Svante. 'Natural Resources and Technology: the Debate about Energy Technology in Eighteenth-century Sweden', *Scandinavian Journal of History* **8** (1983), 83–107.

Lindqvist, Svante. 'Labs in the Woods: The Quantification of Technology during Late Enlightenment', in T. Frängsmyr, J. Heilbron and R. Riders, eds, *The Quantifying Spirit in the 18th Century* (Berkeley: University of California Press, 1990), 291–314.

Lindqvist, Svante. *Technology on Trial: The Introduction of Steam Power Technology into Sweden, 1715–1736* (Stockholm: Almqvist & Wiksell, 1985).

Lindsay, Jack. *Death of the Hero: French Painting from David to Delacroix* (London: Bergamo, 1961).

Linnaeus, Carl. *Flora Lapponica* (Amsterdam, 1737).

Linnaeus, Carl. *Species Plantarum: A Facsimile of the First Edition 1753*, Introduction by W.T. Stearn (London, 1957).

Liscombe, R.W. *William Wilkins, 1778–1839* (Cambridge: Cambridge University Press, 1980).

Lister, W.B.C. *A Bibliography of Murray's Handbooks for Travellers and Biographies of Authors, Editors, Revisers, and Principal Contributors* (Norfolk: Dereham Books, 1993).

Lowe, Lisa. *Critical Terrains: French and British Orientalisms* (Ithaca and London: Cornell University Press, 1991).

Lucas, Colin. 'Great Britain and the Union of Norway and Sweden', *Scandinavian Journal of History* **15** (1990), 269–78.

Lundgren, Anders. 'The New Chemistry in Sweden: The Debate that Wasn't', *Osiris*, 2nd series, **4** (1988), 146–68.

Lyall, Robert. *The Character of the Russians, and a Detailed History of Moscow* (London, 1823).

MacCormack, Carol. 'Medicine and Anthropology', in W. Bynum and R. Porter, eds, *Companion Encyclopaedia of the History of Medicine* (London: Routledge, 1993), 1436–48.

Malthus, Thomas Robert. *An Essay on the Principle of Population, as it affects the Future Improvement of Society, with remarks on the speculations of Mr. Godwin, M. Condorcet, and other writers* (London: J. Johnson, 1798; facsimile reprint, Macmillan, 1966).

Malthus, Thomas Robert. *An Essay on the Principle of Population*, 2 vols; 1803 edition, ed. Patricia James (Cambridge: Cambridge University Press, 1989).

Malthus, Thomas Robert. *Travel Diaries. See* James, Patricia.

Marchand, L., ed. *Byron's Letters and Journals*, 12 vols (London: John Murray, 1973).

Marker, G. *Publishing, Printing, and the Origins of Intellectual Life in Russia, 1700–1800* (Princeton, NJ: Princeton University Press, 1985).

Marsden, William. *History of Sumatra* (1783).

Marshall, Joseph. *Travels through Holland, Flanders, Russia, the Ukraine, Poland . . . in the years 1768, 1769, 1770*, 3 vols (London, 1772).

Marshall, P.J. 'Empire and Authority in the Later Eighteenth Century', *Journal of Imperial and Commonwealth History* **15** (1987), 105–22.

Marshall, P.J. and G. Williams. *The Great Map of Mankind: Perceptions of New Worlds in the Age of Enlightenment* (Cambridge, MA: Harvard University Press, 1982).

McClellan, Andrew. *Inventing the Louvre: Art, Politics, and the Origins of the Modern Museum in Eighteenth-century Paris* (Cambridge: Cambridge University Press, 1994).

McGrew, Roderick E. *Paul I of Russia, 1754–1801* (Oxford: Clarendon Press, 1992).

McKillop, Alan D. 'Local Attachment and Cosmopolitanism – The Eighteenth-Century Pattern', in F.W. Hilles and H. Bloom, eds, *From Sensibility to Romanticism* (New York, 1965), 191–218.

McNeil, Maureen. *Under the Banner of Science: Erasmus Darwin and His Age* (Manchester: Manchester University Press, 1987).

McNeill, William. *Europe's Steppe Frontier, 1500–1800* (Chicago: University of Chicago Press, 1964).

Metcalf, Michael. 'The First "Modern" Party System? Political Parties, Sweden's

Age of Liberty and the Historians', *Scandinavian Journal of History* 2 (1977), 265–87.

Middleton, N. *Travels as a Brussels Scout* (London: Weidenfeld and Nicolson, 1997).

Miller, David and P. Reill, eds. *Visions of Empire: Voyages, Botany, and Representations of Nature* (Cambridge: Cambridge University Press, 1996).

Miller, E. *That Noble Cabinet: A History of the British Museum* (London: André Deutsch, 1973).

Miller, William. *The English in Athens before 1821* (London: Anglo-Hellenic League, 1926).

Millward, Roy. *Scandinavian Lands* (London: Macmillan, 1964).

Monk, J.H. and C. Blomfield, eds. *Museum Criticum; or, Cambridge Classical Researches*, 7 nos (Cambridge, 1814–26).

Moravia, Sergio. 'The Enlightenment and the Sciences of Man', *History of Science* 18 (1980), 247–68.

Morgan, F. 'Between Primates and Primitives: Natural Man as the Missing Link in Rousseau's Second Discourse', *Journal for the History of Ideas* 54 (1993), 37–58.

Moyne, Ernest J. *Raising the Wind: The Legend of Lapland and Finland Wizards in Literature* (Newark: University of Delaware Press, 1981).

Müller, G.F. *Voyages from Asia to America for completing the discoveries of the north-west coast of America*, translated by Thomas Jeffreys (London, 1761).

O'Dell, Andrew C. *The Scandinavian World* (London: Longman, 1956).

Oddy, John. *European Commerce, shewing new and secure Channels of Trade with the Continent of Europe: detailing the Produce, Manufactures, and Commerce, or Russia, Prussia, Sweden, Denmark, and Germany; as well as the Trade of the Rivers Elbe, Weser, and Ems* (London, 1805).

Okenfuss, Max J. *The Rise and Fall of Latin Humanism in Early-modern Russia: Pagan Authors, Ukrainians, and the Resiliency of Muscovy* (Leiden: E.J. Brill, 1995).

Oliva, L. Jay. *Russia in the Era of Peter the Great* (Englewood Cliffs, NJ: Prentice Hall, 1969).

Otter, William. *The Life and Remains of Edward Daniel Clarke*, 2 vols (London: Cowle & Co., 1825).

Ouzouf, Mona. *Festivals of the French Revolution* (Cambridge MA: Harvard University Press, 1988).

Pagden, Anthony. *European Encounters with the New World* (New Haven, CT: Yale University Press, 1993).

Pallas, Peter Simon. *Travels through the Southern Provinces of the Russian Empire, performed in the years 1793 and 1794*, translated by Francis Blagdon in 2 vols for his *Modern Discoveries* (London, 1802–3).

Perry, John. *The State of Russia under the Present Czar, in relation to the several great and remarkable things he has done, as to his naval preparations, the regulating his army, the reforming his people, and improving his countrey* [sic] (London, 1716).

Phillips, Richard. See Goldsmith, J.

Pimlott, J.A.R. *The Englishman's Holiday: A Social History* (London: Faber & Faber, 1947).

Pinkerton, John. *A General Collection of the best and most interesting Voyages and Travels in all parts of the World, many of which are new first transla-tions into English,* 17 vols (London, 1808–14).

Pinkerton, John. *Modern Geography, a description of the Empires, states and colonies, with the oceans, seas, and islands in all parts of the World,* 2 vols (London, 1802).

Pocock, J.G.A. 'Gibbon's *Decline and Fall* and the World View of the Late Enlightenment', in idem, *Virtue, Commerce, and History: Essays on Political Thought and History, Chiefly in the Eighteenth Century* (Cambridge: Cambridge University Press, 1985), 143–56.

Pocock, J.G.A. *The Machiavellian Moment: Florentine Political Thought and the Atlantic Republican Tradition* (Princeton, NJ: Princeton University Press, 1975).

Pocock, J.G.A. *Virtue, Commerce and History: Essays on Political Thought and History* (Cambridge: Cambridge University Press, 1985).

Pontoppidan, Eric. *Natural History of Norway* (originally in Danish, 1751; English translation: London, 1755).

Porter, Dorothy, ed. *The History of Public Health and the Modern State* (Amsterdam: Rodopi, 1994).

Porter, Dorothy. *Health, Civilisation, and the State: The History of Public Health from Antiquity to Modernity* (London: Routledge, 1999).

Porter, Roy. *The Making of Geology: Earth Science in Britain, 1660– 1815* (Cambridge: Cambridge University Press, 1977).

Porter, Roy and Mikulás Teich, eds. *The Enlightenment in National Context* (Cambridge: Cambridge University Press, 1981).

Porter, Roy and Mikulás Teich, eds. *Romanticism in National Context* (Cambridge: Cambridge University Press, 1988).

Porter, T.M. 'The Promotion of Mining and the Advancement of Science: The Chemical Revolution in Mineralogy', *Annals of Science* **38** (1981), 543–70.

Pratt, Mary Louise. *Imperial Eyes: Travel Writing and Transculturation* (London: Routledge, 1992).

Preston, T.R. 'Biblical Criticism, Literature, and the Eighteenth-century Reader,' in I. Rivers, ed., *Books and their Readers in Eighteenth-Century England* (Leicester: Leicester University Press, 1982), 97–126.

Price, Richard. *Observations on Reversionary Payments* (London, 1771).

Prichard, James C. *Researches into the Physical History of Man* (London, 1813).

Putnam, Peter, ed. *Seven Britons in Imperial Russia: 1698–1812* (Princeton, NJ: Princeton University Press, 1952).

Quennell, P. *Byron: A Self-Portrait: Letters and Diaries, 1798 to 1824,* 2 vols (London: John Murray, 1950).

Quintana, Ricardo. *Oliver Goldsmith: A Georgian Study* (London: Macmillan, 1967).

Raeff, Marc. 'The Enlightenment in Russia and Russian Thought in the Enlightenment', in J.G. Garrard, *The Eighteenth Century in Russia* (Oxford: Clarendon Press, 1973), 25–47.

Ragsdale, Hugh, ed. *Paul I: A Reassessment of His Life and Reign* (Pittsburgh, PA: UCIS Series in Russian and East European Studies, 1979).

Ragsdale, Hugh. *Tsar Paul and the Question of Madness: An Essay in History and Psychology* (New York & London: Greenwood Press, 1988).

Rashid, Salim. 'Malthus's Essay on Population: The Facts of "Super-Growth" and the Rhetoric of Scientific Persuasion', *Journal for the History of Behavioural Sciences* **23** (1987), 22–36.

Rice, Tamara Talbot. 'The Conflux of Influences in Eighteenth-Century Russian Art and Architecture: a Journey from the Spiritual to the Realistic', in J.G. Garrard, ed., *The Eighteenth Century in Russia* (Oxford: Clarendon Press, 1973), 267–99.

Richard, John. *A Tour from London to Petersburg and from thence to Moscow* (Dublin, 1781).

Richardson, John. *A Dissertation on the Languages, Literature, and Manners of Eastern Nations* (Oxford, 1777).

Richardson, William. *Anecdotes of the Russian Empire; In a series of Letters written a few years ago from St Petersburg* (London, 1784).

Rigby, Brian. 'Volney's Rationalist Apocalypse: "Les Ruines ou méditations sur les révolutions des empires" ', in F. Barker et al., eds., *1789: Reading, Writing, Revolution* (Proceedings on the Essex conference on the Sociology of Literature: University of Essex, 1982), 22–37.

Ritvo, Harriet. *The Platypus and the Mermaid and other Figments of the Classifying Imagination* (Cambridge, MA: Harvard University Press, 1997).

Rivers, I., ed. *Books and their Readers in Eighteenth-Century England* (Leicester: Leicester University Press, 1982).

Roberts, Michael. *The Age of Liberty: Sweden 1719–1772* (Cambridge: Cambridge University Press, 1986).

Robinson, Jane. *Wayward Women: A Guide to Women Travellers* (Oxford: Oxford University Press, 1990).

Rorlich, 'Azade-Ayse. *The Volga Tatars: A Profile in National Resilience* (Stanford: University of California Press, 1986).

Rosen, Charles. *A History of Public Health* (New York, 1958, new edition: Johns Hopkins, 1993).

Rothenberg, Jacob. *'Descensus ad Terram': The Acquisition and Reception of the Elgin Marbles* (New York and London: Garland Publishing, 1977).

Rowse, A.L. *Four Caroline Portraits* (London: Duckworth, 1993).

Russell, A. 'John Henry Heuland', *Mineralogical Magazine* **29** (1952), 395–405.

Russell, William. *The History of Modern Europe: with an account of the Decline and Fall of the Roman Empire; and a view of the Progress of Society, from the Rise of the Modern Kingdoms, to the Peace of Paris in 1763, in a series of letters from a Nobleman to his Son*, 5 vols (London, 1786–1805).

Ryan, A.N. 'The Defence of British Trade with the Baltic, 1808–1813', *English Historical Review* **74** (1959), 443–66.

Said, Edward. *Orientalism: Western Conceptions of the Orient* (London: Routledge and Kegan Paul, 1978).

Saul, Norman E. 'The Objectives of Paul's Italian Policy', in Hugh Ragsdale, ed., *Paul I: A Reassessment of His Life and Reign* (Pittsburgh, PA: UCIS Series in Russian and East European Studies, 1979), 31–43.

Schroeder, Paul. 'The Collapse of the Second Coalition', *Journal of Modern History* **59** (1987), 244–90.

Scott, H.M. ed. *The European Nobilities in the Seventeenth and Eighteenth Centuries, Vol. II: Northern, Central, and Eastern Europe* (London and New York: Longman, 1995).

Searby, Peter. *A History of the University of Cambridge: Volume 3: 1750–1870* (Cambridge: Cambridge University Press, 1997).

Shaffer, E.S. *'Kubla Khan' and the Fall of Jerusalem: The Mythical School in Biblical Criticism and Secular Literature, 1770–1880* (Cambridge: Cambridge University Press, 1975).

Shaffer, E.S. 'Romantic Philosophy and the Organisation of the Disciplines: the Founding of the Humboldt University of Berlin', in Andrew Cunningham and Nicholas Jardine, eds, *Romanticism and the Sciences* (Cambridge: Cambridge University Press, 1990), 38–54.

Shaw, George. *Musei Leveriani explicatio, Anglica et Latina* (London, 1792).

Shaw, George. *General Zoology, or Systematic Natural History*, 14 vols (London, 1800–26).

Sheridan, Charles Francis. *A History of the Late Revolution in Sweden* (London, 1778).

Sheridan, T. *A Plan of Education for the Young Nobility and Gentry of Great Britain* (London, 1769).

Simmonds, A.T. 'The Founders: The Rt. Hon. Charles Greville, F.R.S., F.L.S. (1749–1809)', *Journal of the Royal Horticultural Society of London* **67** (1942), 219–32.

[Sinclair, John.] *General Observations regarding the Present State of the Russian Empire* (London, 1787).

Smith, Arthur Hamilton. 'Lord Elgin and His Collection', *Journal of Hellenic Studies* **36** (1916), 163–372.

Smith, Roger. *The Fontana History of Human Sciences* (London: Fontana, 1997).

Smith, Samuel Stanhope. *An Essay on the Causes of the Variety of Complexion and Figure in the Human Species* (1787; reprint, ed. Winthrop D. Jordan, Cambridge, MA: Harvard University Press, 1965).

Smith, W.C. 'A History of the First Hundred Years of the Mineral Collection in the British Museum, with Particular Reference to the Work of Charles Konig', *Bulletin of the British Museum* (Natural History) (historical series) **3** (1969), 235–59.

Smollett, Tobias. *Travels through France and Italy* (London, 1766).

Smyth, William. 'A List of Books recommended and referred to in the Lectures on Modern History '(Cambridge, 1815).

Sörlin, Sverker. 'Scientific Travel – the Linnaean Tradition', in T. Frängsmyr, ed., *Science in Sweden: The Royal Swedish Academy of Sciences 1739–1989* (Canton, MA: Science History Publications, 1989), pp. 96–123.

Spate, Virginia, et al. *French Painting: The Revolutionary Decades 1760–1830* (Sydney: Australian Gallery Directors Council, 1980).

St Clair, William. *Lord Elgin and the Marbles* (Oxford, 1967; 2nd edn, Oxford: Oxford University Press, 1983; 3rd edn, Oxford University Press, 1998).

Stepan, Nancy. *The Idea of Race in Science: Great Britain 1800–1960* (London: Macmillan, 1982).

Stone, Lawrence, ed. *The University in Society*, 2 vols (Princeton, NJ: Princeton University Press, 1974).

Stoneman, Richard. *Land of Lost Gods: The Search for Classical Greece* (London: Hutchinson, 1987).

Strahlenberg, P.J. von. *An Historico-Geographical Description of the North and*

Eastern Parts of Europe and Asia; But more particularly of Russia, Siberia, and Great Tartary; Both in their Ancient and Modern State . . . (London, 1736; reprinted 1738).

Stuart, J. and N. Revett, *Antiquities of Athens* (London, Vol. I, 1762; Vol. II, 1788 (though dated 1787); Vol. III (edited by Willey Reveley), 1794; Vol. IV, 1814; Supplement, 1830).

Subtelny, Orest. *A History of Ukraine* (Toronto: University of Toronto Press, 1988).

Sutherland, L.S. and L.G. Mitchell, eds. *The History of the University of Oxford: Volume 5: The Eighteenth Century* (Oxford: Clarendon Press, 1986).

Swinton, Andrew. *Travels into Norway, Denmark, and Russia in the years 1788, 1789, 1790, and 1791* (London, 1792).

Thomas, Keith. *Man and the Natural World: Changing Attitudes in England, 1500–1800* (London: Allen Lane, 1983).

Thomas, Nicholas. *Entangled Objects: Exchange, Material Culture, and Colonialism in the Pacific* (Cambridge, MA and London: Harvard University Press, 1991).

Thomson, Thomas. *Travels through Sweden, during the autumn of 1812* (London, 1813).

Tickell, Thomas. *On the Prospect of Peace* (London, 1713).

Tooke, William. *View of the Russian Empire During the Reign of Catharine* [sic] *the Second*, 3 vols (London, 1799).

Towner, John. 'The Grand Tour: A Key Phase in the History of Tourism', *Annals of Tourism Research* (1985), 297–333.

Treasure, Geoffrey. *The Making of Modern Europe, 1648–1780* (London: Routledge, 1985).

Tregaskis, H. *Beyond the Grand Tour: The Levant Lunatics* (London: Ascent Books, 1979).

Tribe, Keith. *Land, Labour, and Economic Discourse* (London: Routledge and Kegan Paul, 1978).

Trusler, John. *The Habitable World Described, or the Present State of the People in all Parts of the Globe, from North to South*, 9 vols (London, 1788–91).

Turner, Frank. *The Greek Heritage in Victorian Britain* (New Haven and London: Yale University Press, 1981).

Turner, Jane, ed. *The Dictionary of Art*, 34 vols (London: Macmillan, 1996).

Tweddell, Robert. *Remains of John Tweddell, Late Fellow of Trinity College, Cambridge* (London, 1816).

Tyson, Edward. *Orang-Outang; or, the Anatomy of a Pygmy Compared with that of a Monkey, an Ape, and a Man* (London, 1699; second edn, 1751).

Upton, A.F. 'The Swedish Nobility, 1600–1772', in H.M. Scott, ed., *The European Nobilities in the Seventeenth and Eighteenth Centuries, Vol. II: Northern, Central, and Eastern Europe* (London and New York: Longman, 1995).

Vaughan, John. *The English Guide Book, 1780–1870: An Illustrated History* (Newton Abbot: David & Charles, 1974).

Venn, J. *Biographical History of Gonville and Caius College*, 7 vols (Cambridge, 1897–1901).

Volney, C.F. *Les Ruines; ou méditations sur les révolutions des empires* (Paris, 1791); translated as *The Ruins: or, A Survey of the Revolutions of Empire* (London, 1792).

Vrettos, Theodore. *A Shadow of Magnitude: The Acquisition of the Elgin Marbles* (New York: Putnam, 1974).

Vrettos, Theodore. *The Elgin Affair: The Abduction of Antiquity's Greatest Treasures and the Passions it Aroused* (London: Secker & Warburg, 1997).

Walker, J.S. *An Essay on the Education of the People* (London, 1825).

Walpole, Robert, ed. *Travels in Various Countries of the East; Being a Continuation of Memoirs relating to European and Asiatic Turkey* (London: Longman, Hurst, Rees, Orme, and Brown, 1820).

Walpole, Robert. *Memoirs relating to European and Asiatic Turkey, edited from manuscript journals*, 2 vols (London, 1817).

Watkin, David. *Sir John Soane: Enlightenment Thought and the Royal Academy Lectures* (Cambridge: Cambridge University Press, 1996).

Watkin, David. *Thomas Hope, 1769–1831, and the New Classical Idea* (London: John Murray, 1968).

Watson, J.S. *The Life of Richard Porson, MA, Professor of Greek in the University of Cambridge from 1792 to 1808* (London, 1861).

Weber, F.C. *The Present State of Russia; Being the journal of a foreign minister who resided in Russia at that time, translated from the Dutch*, 2 vols (London, 1722–23).

Williams, G. *The Great South Sea: English Voyages and Encounters, 1570–1750* (New Haven and London: Yale University Press, 1997).

Williams, John. *The Rise, Progress, and Present State of the Northern Governments; viz., the United Provinces, Denmark, Sweden, Russia, and Poland*, 2 vols (London, 1777).

Wilson, Francesca. *Muscovy: Russia through Foreign Eyes, 1553–1900* (London: George Allen & Unwin, 1970).

Winckelmann, J.J. *The History of Ancient Art Among the Greeks*, trans. G. Henry Lodge (London, 1850).

Winstanely, D.A. *The University of Cambridge in the Eighteenth Century* (Cambridge, 1922).

Wokler, R. 'Anthropology and Conjectural History in the Enlightenment', in Christopher Fox, Roy Porter and Robert Wokler, eds, *Inventing Human Science: Eighteenth-Century Domains* (Berkeley and London: University of California Press, 1995).

Wolff, Larry. *Inventing Eastern Europe: The Map of Civilization on the Mind of the Enlightenment* (Stanford: Stanford University Press, 1994).

Wollstonecraft, Mary. *Letters written during a short residence in Sweden, Norway, and Denmark* (1796), ed. C.H. Poston (Lincoln and London: University of Nebraska Press, 1976).

Woodhouse, C.M. *The Philhellenes* (London: Hodder and Stoughton, 1969).

Woolf, Stuart. 'The Construction of a European World-View in the Revolutionary-Napoleonic Years', *Past & Present* **137** (1992), 72–101.

Wraxall, Nathaniel. *A Tour through some of the Northern parts of Europe, particularly Copenhagen, Stockholm, and Petersburg, in a series of letters*, 2nd edn (London, 1775).

Wright, Herbert G. 'Defoe's Writings on Sweden, '*The Review of English Studies* **16** (1940), 25–32.

Wright, J.M.F. *Alma Mater: or, Seven Years at the University of Cambridge*, 2 vols (London: Black, Young & Young, 1827).

Young, Brian. *Religion and Enlightenment in Eighteenth-century England: Theological Debate from Locke to Burke* (Oxford: Clarendon Press, 1998).

Young, Robert M. 'Malthus and the Evolutionists: the Common Context of Biological and Social Theory', in idem, *Darwin's Metaphor: Nature's Place in Victorian Culture* (Cambridge: Cambridge University Press, 1985), 23–55.

Ziolkowski, T. *German Romanticism and its Institutions* (Princeton, NJ: Princeton University Press, 1990).

Zippel, Otto. *Thomson's Seasons: Critical Edition* (Berlin: Mayer & Müller, 1908).

Index